W9-AKC-488

PETROCHEMICALS

in NonTechnical
LANGUAGE

Third Edition

PETROCHEMICALS

in NonTechnical
LANGUAGE

Third Edition

Donald L. Burdick

William L. Leffler

Copyright© 2001 by
PennWell Publishing Company
1421 South Sheridan/P.O. Box 1260
Tulsa, Oklahoma 74101
1-800-752-9764
sales@pennwell.com
www.pennwell.com
www.pennwell-store.com

Printed in the United States of America

04   03   02   01   5   4   3   2   1

Library of Congress-Cataloging-in-Publication Data

Burdick, Donald L.
   Petrochemicals in nontechnical language / Donald L. Burdick,
William L. Leffler ; [edited by Sue Rhodes Dodd].-- 3rd ed.
      p. cm.
   Includes index.
   ISBN 0-87814-798-5
   1. Petroleum chemicals.   I. Leffler, William L. II. Dodd, Sue
Rhodes. III. Title.
TP692.3 .B873 2001
661'.804--dc21

                                                    2001033230

Edited by Sue Rhodes Dodd, Amethyst Enterprises, Tulsa, Oklahoma.
Managing Editor: Marla M. Patterson
Designed by Amy Spehar

v

CONTENTS ◇

PREFACE ⬡

> "But now ask the beasts,
> and they shall teach thee."
>
> **Job 12:7**

We have updated, edited, and rewritten every chapter in this book and even added two more chapters, qualifying us to use the advertisers' mantra, "New and Improved." After 10 years in the market place, the need to create a third edition came to us like a paper cut from licking an envelope. We were rereading parts of the 1990 edition — the chapters on polymers — and noted we said that a big market for polyvinyl chloride was phonograph records. Later on we said, "probably all the 'wood' on the front of your console TV is polystyrene." Well, lifestyles change with time and so do technologies, stimulating us to produce a new, more useful edition.

After you buy this book, you can use it in at least five ways (besides unabashedly displaying it on your office bookshelf):

- Read it cover to cover for a nontechnical education covering 90% (by volume) of the traded petrochemicals. There are even exercises at the end of each chapter to test comprehension and retention. Complete answers are in the back of the book.
- Read a chapter or section as the subjects come up in your business life. Each one is designed to be a self-contained description of one petrochemical. If you're too busy, there's a 10-sentence summary at the end of each chapter.

- Use it as a nontechnical encyclopedia. The glossary in the back has nearly 300 technical terms and is blessedly nontechnical. And if you can't find what you need in the index at the very end of the book, you should have gone to engineering school because you now have a job where you're in over your head.
- Use it as a primer in petrochemical economics. Many of the chapters have material balances, and a number of the exercises deal with product or process economics.
- Recommend it to your subordinates, colleagues, or your superiors who need to know at least half as much about petrochemicals as you do.

There are four parts to this book, if you leave out the housekeeping and appendices. The first is only one chapter — the mandatory discussion of chemistry. Our editors tell us the book would not be technically complete without it. It's not bad, but we met a reader once who just skimmed it and did "okay" with the rest.

The next part covers the building blocks, from which most of the remaining petrochemicals are derived. The third part, a large midsection, has all the first and second line derivatives.

The next section is on polymers, which are "borderline" petrochemicals. We debated whether they belong in a book about petrochemicals, but then we wrote them and they seemed to complete the linkage from raw materials (coal, oil, gas) all the way to consumer products. If you don't agree, don't read them.

Finally, come the quick references, the Glossary and the Index. Use them when you have no time for pedantic endeavor.

A note about nomenclature:

We switch back and forth indiscriminately between synonyms and different conventions in this book. For example, we use butylene sometimes and butene others; C_2H_4 sometimes, $CH_2=CH_2$ others; iso-butane and isobutane. That's the way it is in industry so you might as well get used to it here.

ACKNOWLEDGEMENTS

In the first and second editions of this book, we never gave appropriate recognition to the infrangible license and ineffable support given to us by our wives, JoAnn and Eileen. Now is the time, we have heard, to remedy that. Thank you.

D. L. B.
W. L. L.

CHAPTER 1 ◌

What You Need to Know about Organic Chemistry

"The time has come," the Walrus
said, "to speak of many things:
of shoes—and ships—and sealing wax—
of cabbages—and kings."

Through the Looking Glass,
Lewis Carroll, 1832–1898

What is organic chemistry? It's the study of compounds containing carbon, and it's fundamental to understanding petrochemicals. Why the word *organic,* you might ask. Originally, and that means before 1800, organic was applied only to compounds whose formation was supposed to be due to some living force such as plants or animals. Then early in the 19th century, a chemist named Wohler synthesized urea, the main ingredient in urine. (Goodness knows why he was trying to do that.) Up until that time it was believed urea could only be produced "organically" by animal life. Therefore, and until today, the term *organic chemistry* was stretched beyond its original meaning to include all carbon compounds. So now the difference between organic and inorganic chemistry is more definitional than natural.

You may be surprised to find out that organic compounds comprise more than 95% of all compounds known to exist, and that's more than a million. Three things about carbon and carbon compounds help explain the proliferation of organic chemicals. The first is the electronic configuration of the carbon element. Don't leave now—you're about to get the six-minute summary of the periodic table of elements, atoms, electrons, protons, valences, bonds, and compounds.

About 100 different kinds of atoms make up all kinds of matter, and they are classified in a table—the Periodic Table of Elements—according to their construction. The center of any atom is a nucleus containing protons and neutrons. The protons have a positive charge and the neutrons are neutral; so the nucleus is positively charged. Electrons, equal in number but opposite in charge to the protons, move around the nucleus in orbits. You might think of an atom like a solar system. The nucleus acts like the sun; the electrons orbit the nucleus like the planets circle the sun.

There is one difference, however. The innermost orbit can contain either one or two electrons, at most. The next orbit can have up to eight electrons. The succeeding orbits become a more complex story, but luckily the atoms that make up almost all petrochemicals have no more than two orbits.

The rules of electrons and orbits are important because the number of electrons in the outermost ring determines some of the more important chemical properties of that atom or element. Atoms have a yearning to move toward maximum stability by filling up their outermost orbit to the maximum content. Atoms can gain or shed electrons or share them with another atom in the process of achieving the stability of a complete set, again, of two or eight electrons for most petrochemicals.

For example, take the carbon atom. It has six neutrons and six protons in the nucleus and six electrons in orbit. The first orbit has two; the second has the four it needs to balance off the four protons. These four are called the valence electrons. Carbon has a valence of four because it needs four more electrons to fill the outer ring up to its capacity of eight. It desperately wants to find some other atoms with which it can share four electrons.

Another example is hydrogen. Hydrogen has one proton and one neutron in the nucleus and one electron in the first and only orbit. It needs another electron in that orbit to stabilize itself. Figure 1–1 shows how carbon and hydrogen can achieve mutual satisfaction in the marriage of two atoms into a compound, methane.

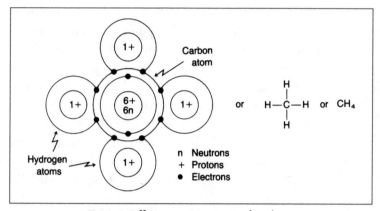

Fig. 1–1 Different representations of methane

Each of the four hydrogen atoms shares its one electron with the carbon atom to satisfy carbon's need for eight electrons in its outer ring. The carbon shares an electron with each hydrogen atom to satisfy the need for two electrons in the hydrogen atoms' outer ring. The polygamous result is a stable compound with all the proton and electron charges balanced.

Carbon and hydrogen can link up with other carbons and hydrogens. When hydrogen hooks up with another hydrogen, it forms H_2, and the electron "urges" of the two hydrogen atoms are satisfied. But when carbon hooks up with another carbon, each carbon still has a need for three more electrons. Filling them out with hydrogens is possible, and when that happens, the compound *ethane* forms, as shown in Figure 1–2.

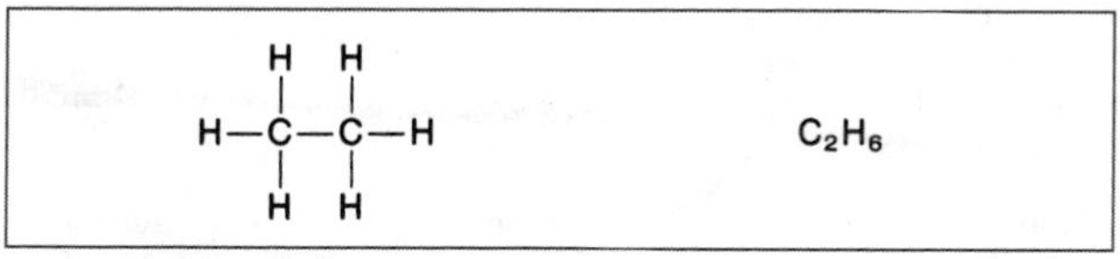

$$H-\overset{\displaystyle H}{\underset{\displaystyle H}{C}}-\overset{\displaystyle H}{\underset{\displaystyle H}{C}}-H \qquad\qquad C_2H_6$$

Fig. 1–2 Ethane

This long narrative about the propensity of carbon to connect with four other atoms partially explains why there are so many carbon compounds: there are lots of ways atoms can hook up with carbon atoms.

A second characteristic unique to carbon compounds, *isomerism*, also helps explain its abundance in nature. Compounds with the same number and kinds of atoms can have very different properties. For example, glucose has the formula $C_6H_{12}O_6$. Yet there are 16 other compounds with the same number of carbons, hydrogens, and oxygens. It's not likely, though, that you'd like your night nurse to hook up galactose or fructose to your intravenous instead of glucose, even thought they have the same formula. The difference is that the atoms are linked together in such a way as to have different spatial configurations, and, as you'll see, that makes them behave differently, both physically and chemically. Such similar but different compounds are called *isomers*.

So, if you put the phenomenon of isomerism together with the propensity of carbon to react (the valence of four), and add to that nature's bountiful supply of carbon on this planet, you can understand the preponderance of compounds and the importance of organic chemistry.

One further characteristic unique to carbon is important and needs to be covered before leaving the subject of valences: *bonds*. A few paragraphs ago, you saw that carbon could link up to itself and three other atoms. In fact, carbon can also link up to itself with double bonds or triple bonds to "satisfy" its valence requirements of four. For example, in Figure 1–3. two carbon atoms are linked together with single, double, or triple bonds filled out with hydrogens, forming three different compounds: ethane, ethylene, and ethyne, or as it's more commonly known, *acetylene*.

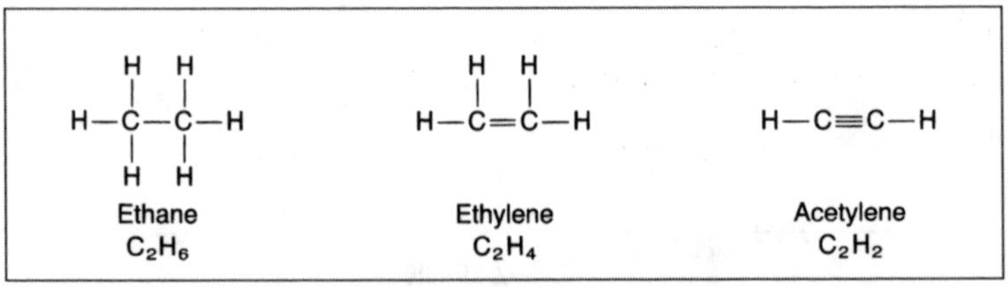

Fig. 1–3 Ethane, ethylene, and acetylene

You will find that in petrochemical processes, the more multiple bonds, the more unstable the compound is, meaning it is likely to engage in a chemical reaction to change its urge to fill up its rings. Acetylene is much more likely to react with other compounds, explosively sometimes, than ethylene, which itself is far more reactive than ethane. You can think of the double or triple bonds as squeezing into the place suitable for one bond. As a result, there is pent-up pressure to relieve the stress in the form of increased chemical reactivity.

As a matter of common nomenclature in the petrochemical world (at least when you hear chemical engineers or chemists talking), carbon compounds with single bonds are sometimes called *saturates*. (The carbon atoms are saturated with other atoms.) Those with multiple bonds are called *unsaturates*.

Double bonds characterize the basic building blocks of the petrochemical business. Ethylene, for example, is the chemical compound used to make vinyl chloride, ethylene oxide, acetaldehyde, ethyl alcohol, styrene, alpha olefins, and polyethylene, to name only a few. Propylene and benzene, the other big-volume building blocks, also have the characteristic double bonds.

Without a road map, going on from here can be a tangled web. You need to look at one of the generally accepted breakdowns of organic chemicals, shown in Figure 1–4.

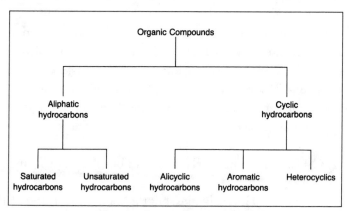

Fig. 1–4 Classification of organic chemicals

You have already seen the aliphatic hydrocarbons. They contain only hydrogen and carbon atoms; they can have single or multiple bonds. (The word *aliphatic* has Greek origin with the meaning *coming from fat*, or more loosely, *coming from an organic compound.* This is another remnant of medieval science.)

The simplest member of the aliphatic group is methane. More complicated molecules (combinations of atoms) in this group can be formed by adding additional combinations of one carbon with two hydrogens attached to it between any carbon and hydrogen atoms, as in Figure 1–5.

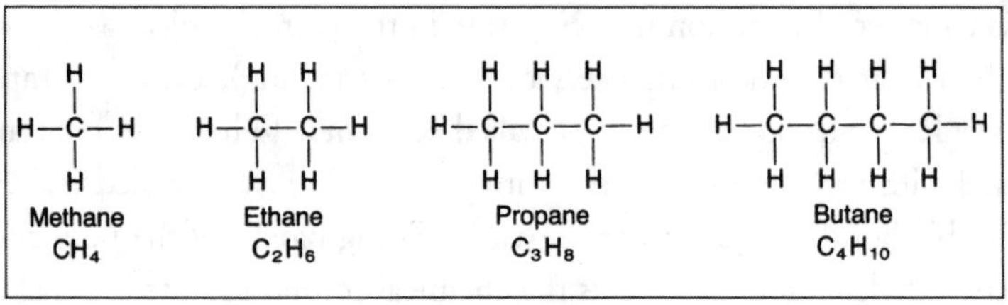

Fig. 1–5 Paraffins

The whole family that results from endless addition of the -CH$_2$'s is called *the paraffin series*. The word comes from the name of the wax little old ladies used in "the old days" to seal jelly jars. That particular paraffin consists of a mixture of C$_{30}$H$_{62}$'s on up to C$_{50}$H$_{102}$'s. Note that the formulas always have twice as many hydrogens plus two, compared to the carbons. That's the way it works out.

The petroleum products processed in oil refineries are predominantly paraffins and are often characterized by the temperature at which they boil. Distillation or fractionation, one of the most useful processes in refining, is based on these boiling points. For example, at room temperatures, the following petroleum-type paraffins take the three basic forms of matter we see in nature—gas, liquid, and solid:

CH$_4$, C$_2$H$_6$, C$_3$H$_8$, and C$_4$H$_{10}$ are gases but they liquefy at or below 32°F (0°C)

C$_6$H$_{14}$ through C$_9$H$_{20}$ are liquids, but they boil between 150 and 300°F (65°C and 150°C)

C$_{30}$H$_{62}$ and bigger are solids, but they melt at 200–300°F and boil above 500°F (250°C)

Unsaturated hydrocarbons are typified by ethylene. Propylene, butylene, and bigger molecules are structured in the same manner as the saturates, but one of the single bonds is replaced with a double bond, as shown in Figure 1–6. Another popular name for these compounds is *olefins*.

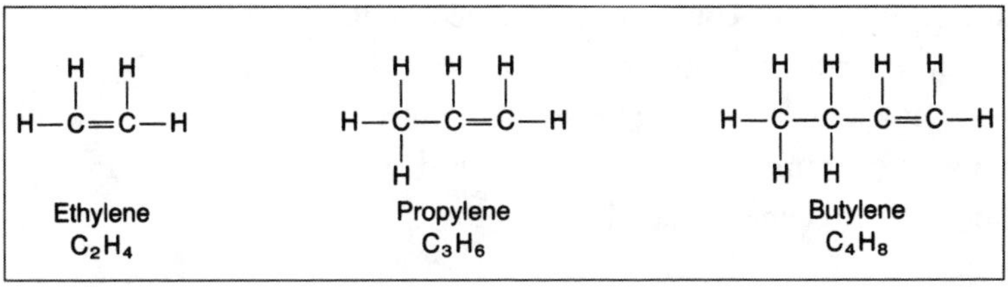

Ethylene
C_2H_4

Propylene
C_3H_6

Butylene
C_4H_8

Fig. 1–6 Olefins

The double bond difference between the olefins and the paraffins is the quintessential difference between the petrochemicals and petroleum products—the petrochemicals industry depends much more on the chemical reactivity of the double-bonded molecules. While paraffins can be manipulated in refineries by separation or reshaping, olefins in a petrochemical plant are usually "reacted" with other organic compounds or another kind of atom or compound such as oxygen, chlorine, water, ammonia, or more of itself. The results are more complicated compounds useful in an increasing number of chemical applications. More on this in later chapters.

Stop for a little interlude to pick up two auxiliary, but important, concepts. The first is the *organic* group. The other is *isomer*, which should have a familiar ring.

Organic group is a handy chemical shorthand notation for a cluster of atoms that looks much like the stand-alone molecule after which it is named. Take the methyl group. It's nothing more than methane with one of the hydrogens missing, as shown in Figure 1–7. But it is attached to some other atoms to make up a larger molecule, *methyl alcohol*. Organic groups are not stand-alone molecules themselves. They are always part of a molecule.[1]

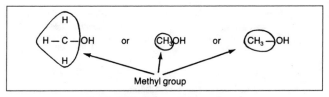

Fig. 1–7 Methyl group in methanol

The shorthand symbol for organic groups is R-. Technical writers (and chemistry teachers) use R- whenever they want to indicate that any number of organic groups could be attached to make a molecule. In Figure 1–7, the methyl group, $-CH_3$, could be represented by R-.

Another good example of organic groups is shown in Figure 1–8, a diagram of tetraethyl lead. This is the additive that was put in gasoline to improve the octane rating. Tetraethyl lead has four ethyl groups ($-C_2H_5$) attached to the element lead (Pb).

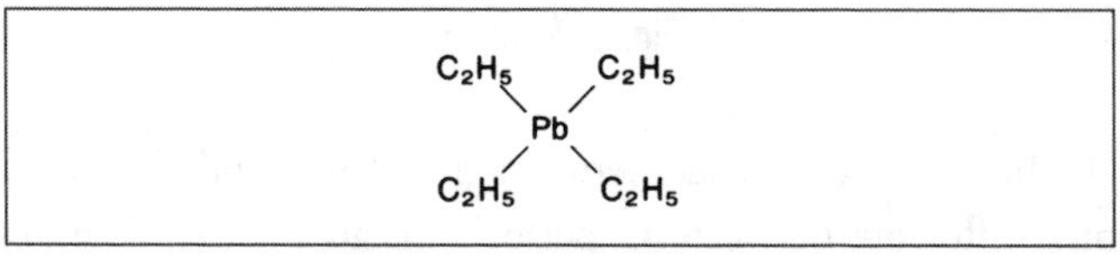

$$C_2H_5 \diagdown \qquad \diagup C_2H_5$$
$$Pb$$
$$C_2H_5 \diagup \qquad \diagdown C_2H_5$$

Fig. 1–8 Tetraethyl lead

For the most part, in the rest of this book, the organic group notations in the middle and at the right side of Figure 1–7 will be used. They are a lot less clumsy than the sprawl at the left. Occasionally, the notation R- will turn up too.

Now back to isomers. To firmly implant them in your mind, consider butane and its isomer, isobutane, in Figure 1–9. The difference between the two C_4H_{10} molecules is how the organic groups $-CH_3$ are connected. In isobutane, one of the carbons has three (not two) methyl groups attached to it. There is more to the difference between the two molecules than just drawing them. Isobutane behaves differently as well. It boils at a different temperature, it gives off a different amount of heat when it burns, it has different chemical reactivity, and so on.

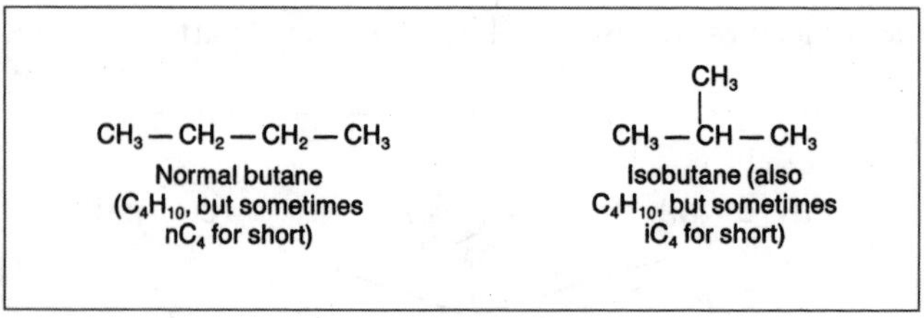

$$CH_3-CH_2-CH_2-CH_3$$
Normal butane
(C_4H_{10}, but sometimes
nC_4 for short)

$$CH_3-\overset{\overset{\textstyle CH_3}{\textstyle |}}{CH}-CH_3$$
Isobutane (also
C_4H_{10}, but sometimes
iC_4 for short)

Fig. 1–9 Butanes

The butylene isomers shown in Figure 1–10 add another degree of complexity because of the double bond. It is an easy mistake to go overboard in drawing isomers that have the same formula but appear to look different. But be careful, because molecules don't know left from right or front from back. What may look different on paper may be identical when rolled over in space. That's why the isobutylene in Figure 1–10 is drawn the way it is. If you try to attach that =CH_2 group to some other carbon in the molecule, the whole thing becomes a normal butylene.

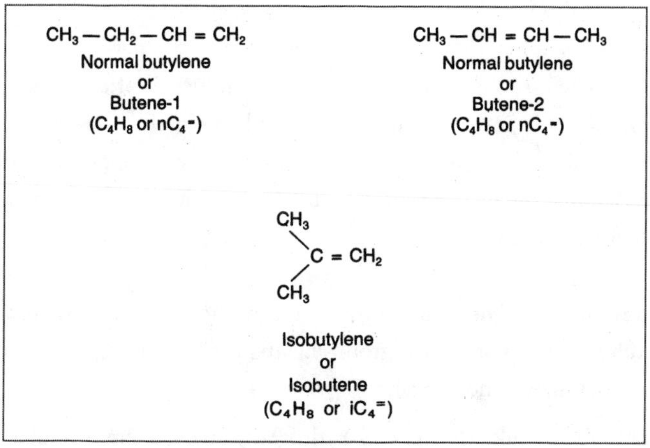

$CH_3 - CH_2 - CH = CH_2$
Normal butylene
or
Butene-1
(C_4H_8 or $nC_4{}^=$)

$CH_3 - CH = CH - CH_3$
Normal butylene
or
Butene-2
(C_4H_8 or $nC_4{}^=$)

CH_3
$\diagdown$
$C = CH_2$
$\diagup$
CH_3

Isobutylene
or
Isobutene
(C_4H_8 or $iC_4{}^=$)

Fig. 1–10 Butylenes

Like the butane isomers, the butylenes each have their own properties that make them unique and of individual appeal to the petrochemical industry.

The root-*mer* figures importantly in petrochemical nomenclature. It comes from the Greek word *meros*, which means *part*. The chemists picked up its usage to define how organic groups are linked together. You will find it imbedded in the following:

monomer (with *mono-*, one)—a compound capable of reacting with itself or other similar compounds, e.g., ethylene.

dimer (with *di-*, two)—two monomers joined together, e.g., butene.

trimer (with *tri-*, three)—three monomers joined together, e.g. hexene.

oligomer   (with *olig-*, a few)—up to 10, more or less, monomers joined together in a string, e.g., alpha olefins.

polymer   (with *poly-*, many)—multiple monomers linked together, e.g., polyethylene.

isomer   (with *iso-*, equal)—molecules with an equal number and kind of atoms arranged differently e.g., butylene and isobutylene.

Dimers, trimers, and **tetramers** are all forms of **oligomers.**

The chemists in the petrochemical industry often characterize compounds with nomenclature starting with the prefix, *alk-*, and ending in a code that helps them remember what's in the compound. It doesn't help much that the plant engineers often corrupt the category names when they name products.

Alkanes are straight-chained or even branch-chained hydrocarbon molecules made up of methyl groups having the formula C_nH_{2n+2}, such as butane, isobutane, and pentane.

Alkenes are like alkanes less two hydrogens, giving them a double bond somewhere in the chain, making them an olefin. They have the formula, C_nH_{2n}, such as butene (more often called butylene), isobutene (isobutylene), and propene (usually called propylene).

Alkynes, moving down the alkane and alkene protocol, and with the formula, C_nH_{2n-2}, have a triple bond between two carbons, such as ethyne, $CH{\equiv}CH$ or butyne, $CH_3-C{\equiv}C-CH_3$. Confusing the protocol are the *dienes*, which also have formula, C_nH_{2n-2}, but have two double bonds in their molecules such as butadiene ($CH_2=CH-CH=CH_2$) or *pentadiene*, but hardly anyone uses the archaic word *alkadiene*.

Alkyls are paraffinic hydrocarbon groups (but not stand-alone compounds) derived from alkanes by dropping one hydrogen from the formula, resulting in $-C_nH_{2n+1}$, such as the ethyl or propyl groups.

> When the suffixes *-ane* and *-ene* show up in a compound name like decane or decene they are usually consistent with the **alkane** and **alkene** definitions, but not always. Benzene is a cyclic aromatic hydrocarbon, not a straight-chain molecule; naphthenes are cyclic compounds.

CYCLIC COMPOUNDS

The fundamental difference between cyclic hydrocarbon compounds and the others already covered is the arrangement of the carbon atoms in a cyclic structure. Cyclic compounds have a closed chain of carbon atoms. Cyclopropane, shown in Figure 1–11, is the simplest cyclic hydrocarbon.

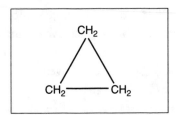

Fig. 1–11 Cyclopropane

Dentists used to administer cyclopropane to their patients—it's one of the several anesthetics used to put you to sleep. Others are nitrous oxide (laughing gas) and ether, which you will run into later on in this book.

Cyclopropane and the cyclic compounds shown in Figure 1–12, cyclopentane and cyclohexane, are members of the alicyclic branch shown in Figure 1–4. The *ali-* is the same prefix as used in the aliphatics because of the structure. Except for the cyclic formation, they are made up basically of chains of methylene groups (-CH₂-). But one difference from the aliphatic series of organics is the chemical reactivity. Lower members of the alicyclic series have one chemical property similar to double-bonded olefins—they are quick to react chemically.

The simple explanation for this reactivity is that the bonds attaching the carbons to each other are strained because of the angles they must take.

They're bent out of their "natural" shape. In any chemical reaction, the rings readily open up to alleviate this strain. The explosive nature of cyclopropane must have made patients—and even their dentists—a little anxious while they were sitting in a dentist's chair with their lungs full of it.

Fig. 1—12 Alicyclics

You might surmise that if there are more carbon atoms in the ring, the compounds might be more stable. In fact, cyclopentane and cyclohexane are much more stable than cyclopropane, and like the paraffins, slower to react. They'll burn easily enough but not explosively.

Cyclopentane and cyclohexane are commonly found in petroleum products like gasoline and are generically called yet another name, *naphthenes*, in the refining business.

Aromatics Compounds

By far, the most commercially important family of compounds on the cyclic side of the roadmap in Figure 1—4 are the aromatic compounds. Benzene is the patriarch. Like much of the nomenclature in organic chemistry, the term *aromatic* is a misnomer. It is a legacy from the 19th century when a group of unsaturated compounds of high reactivity and with a sickly, sweet, hydrocarbonish smell were isolated and fell under the name *aromatics*. Unlike the term *organic*, which got broader meaning in the 20th century, the name *aromatics* got narrower and is limited today to benzene and benzene derivatives.

The benzene molecule is a remarkable structure with six carbons in a hexagonal ring. To satisfy the carbon valence of four, every other carbon-to-carbon link is a double bond, and each carbon has only one hydrogen attached. (*See* Fig. 1–13.)

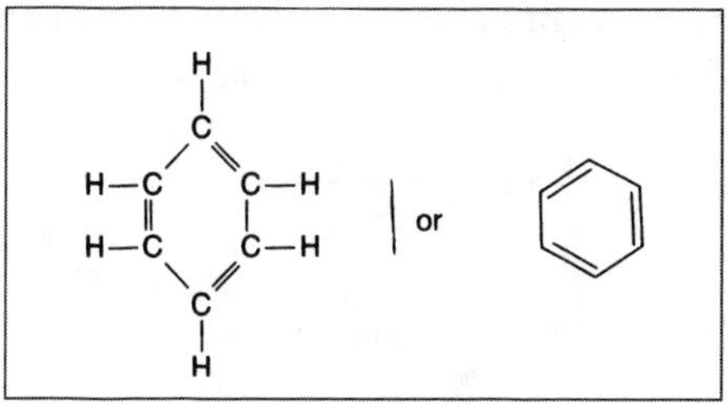

Fig. 1–13 Benzene, C_6H_6

There are some subtle but very important characteristics unique to the benzene ring. One is symmetry. Every carbon in the ring looks like every other carbon; every hydrogen looks like every other hydrogen. There are no benzene isomers. Every benzene molecule looks like every other benzene molecule. Moreover, as in many chemical reactions covered in later chapters, when one of the hydrogens is replaced during a chemical reaction they result in something called *a monosubstituted benzene*, that compound also has no isomers. The beneficial fallout of this phenomenon is that the products of monosubstitution are identical, *homogeneous*.

Take the compound *toluene*, for example. In Figure 1–14, each of the three molecules is benzene with a methyl group replacing a hydrogen. While they appear to be different, each needs to be rotated just a little to make it look like the others. So there's really only one kind of toluene, as toluene is a monosubstituted benzene.

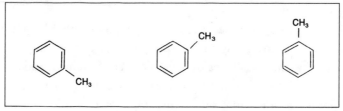

Fig. 1–14 Toluene $C_6H_5CH_3$

The next logical step is to replace two hydrogens on the benzene ring (di-substitution). Three isomers occur. Take the compound xylene, C_6H_4 $(CH_3)_2$, please. That's a benzene ring with methyl groups replacing two hydrogens. As you can see in Figure 1–15, the replacement can be in one of three patterns (and only three, if you look closely).

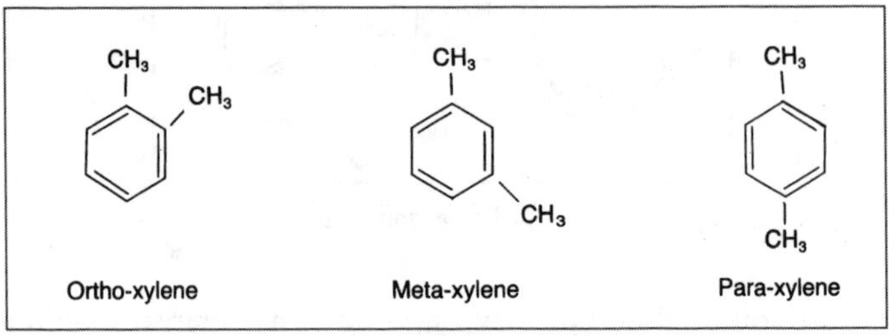

Fig. 1–15 Xylenes

The isomers, called *ortho-xylene*, *meta-xylene*, and *para-xylene*, each have unique properties. Two such properties are the freeze points, at which xylenes turn from liquid to crystals, and the boiling points, at which xylenes turn from liquid to vapor. These two properties figure importantly in the apparatus used to separate xylene isomers from each other. Mixed xylenes, a commonly traded commodity, is a combination of the three isomers.

Di-substituted benzenes like xylenes might be compared to a litter of puppies. They're all dogs, but each one behaves differently. If they were mono-substituted benzenes like toluene, they'd all be clones with the same DNA.

HETEROCYCLICS

Take one of the alicyclic or aromatic compounds that have a chain of carbon atoms in a closed ring and replace one of the carbon atoms with some other kind of atom (usually it's oxygen, nitrogen, or sulfur), and you have a heterocyclic compound. Ethylene oxide is the simplest of the heterocyclic series since it is a three-atom ring. (Anything smaller wouldn't be a ring.) In the ethylene oxide molecule shown in Figure 1–16, the oxygen atom doesn't have any hydrogen atoms attached because it has a valence of two. The marriage of the oxygen and carbon atoms bonds mutually satisfies the valence requirements of each.

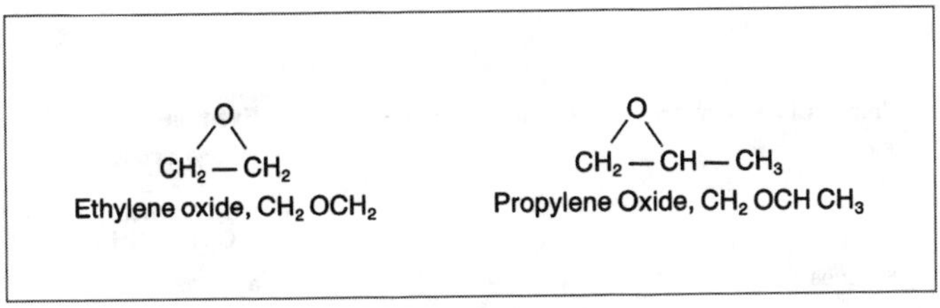

Fig. 1–16 Cyclic oxides

Propylene oxide, another commercially important chemical, also is shown in Figure 1–16. It illustrates two concepts already discussed—heterocyclics and a methyl group replacing a hydrogen atom.

EVERYTHING ELSE

The list of "everything else" is expanding endlessly and could be the longest section of this chapter. University students take numerous advanced courses to learn about them. Mercifully, this section of "everything else" is limited to brief discussion of the few classes of compounds that have become important in the petrochemical business.

You are now off the Figure 1–4 road map and onto making organic compounds by adding something other than carbon and hydrogen atoms. The idea is the same as heterocyclics in Figure 1–4. And as in Figure 1–16, one of the most important of these elements is oxygen.

Oxygenated organic compounds

Table 1–1 shows seven types of organic compounds that have gone through an oxidation process of some kind. They have had oxygen chemically added to them in some fashion. Each of these groups of compounds will be discussed in detail in one or more chapters later on, but some familiarity with the nomenclature at this point is helpful. (In each of the chemical formulas in the table, the letter R is meant to represent some organic group or compound in the manner explained earlier. More importantly, attached to the R is the signature of a family group.)

Family Classification	Generic Formula	Examples
Alcohols	$R - OH$	methyl alcohol $CH_3 - OH$ ethyl alcohol $C_2H_5 - OH$
Ketones	$R - \underset{\underset{O}{\parallel}}{C} - R$	acetone $CH_3 - \underset{\underset{O}{\parallel}}{C} - CH_3$
Aldehydes	$R - \underset{\underset{H}{\mid}}{C} = O$	acetaldehyde $CH_3 - \underset{\underset{H}{\mid}}{C} = O$
Acids	$R - \underset{\underset{O}{\parallel}}{C} - OH$	acetic acid $CH_3 - \underset{\underset{O}{\parallel}}{C} - OH$
Esters	$R - \underset{\underset{O}{\parallel}}{C} - O - R'$	ethyl acetate $CH_3 - \underset{\underset{O}{\parallel}}{C} - OC_2H_5$
Ethers	$R - O - R$	dimethyl ether $CH_3 - O - CH_3$
Anhydrides	$\begin{matrix} R - \overset{\overset{O}{\parallel}}{C} \searrow \\ \quad\quad O \\ R - \underset{\underset{O}{\parallel}}{C} \nearrow \end{matrix}$	acetic anhydride $\begin{matrix} CH_3 - \overset{\overset{O}{\parallel}}{C} \searrow \\ \quad\quad\quad O \\ CH_3 - \underset{\underset{O}{\parallel}}{C} \nearrow \end{matrix}$

Table 1–1 Oxygenated hydrocarbons

The characteristic signature group of alcohols is the addition of the hydroxyl group -OH to another group. An -OH group and a methyl group make methyl alcohol.

The ketones have an imbedded signature, a carbon atom with a double-bonded oxygen attached. Acetone (finger nail polish remover) is the simplest and most common ketone.

Aldehydes have a tail end consisting of a carbon/double-bonded oxygen and a hydrogen, both attached to the same carbon. The commodity, *acetaldehyde*, is a big-volume aldehyde. A better known one is *formaldehyde*.

The acid signature is just a bit more complicated. It's a double-bonded oxygen plus a hydroxyl (-OH) group, both attached to the same carbon. The main ingredient in vinegar, acetic acid, is an example.

Ethers are simple. They have an imbedded oxygen connecting two organic groups that may or may not be identical. Diethyl ether is the one they give you just before they take your appendix out.

Esters get more complicated, having a carbon with single- and double-bonded oxygens. Anhydrides defy simple explanation, so just look at Table 1–1. The most common ester you have probably encountered is methyl acetate, the solvent put in cans of fast-drying spray paint. There aren't any commonly used anhydrides around the house.

Nitrogen-based organic hydrocarbons

Just outside the family of organic compounds (like next-door neighbors) is a family of compounds based on nitrogen. There are three main branches of the family, shown in Table 1–2. The amines are predominant and generally are formed by reactions involving ammonia, NH_3. That's where the *am-* in amine comes from, *ammonia*. Usually the amines have in them organic groups commonly found in the petrochemical industry. (That's what happens in neighborhoods.) Aniline ($C_6H_5NH_2$), a typical example, is an important dye intermediate in which you can find a group derived from benzene.

Family Classification	Generic Formula	Examples
Amines	$R-NH_2$	Aniline
Nitro compounds	$R-NO_2$	Tri-nitro toluene
Nitriles	$R-CN$	Acrylonitrile $CH_2 = CH — CN$

Table 1–2 Nitrogen-based organic compounds

The nitro compounds are organic compounds linked with the grouping -NO_2. Usually the -NO_2 comes from nitric acid, HNO_3, as in the reaction of nitric acid and toluene to make 2,4,6 tri-nitro-toluene, which is TNT. The numbers are a code indicating where the three nitro groups are attached in relation to the methyl group in toluene.

Finally, a family with a slim tree of near-petrochemicals are the *nitriles*, compounds with the signature -CN. The family success in this house is acrylonitrile, a compound used extensively in the manufacture of tires, plastics, and the kind of fibers that go into sweaters (Orlon and Acrylon).

That's it! This is, by far, the toughest chapter in the book. There will be more chemistry as you go along in each chapter, but the doses will be small and easy to swallow. So, if you've gotten this far, engage a plant engineer in conversation and blow him away with a few "alkyls" and "methyl groups." But make it a short conversation.

EXERCISES

1. Match the items in the left column with the correct corresponding items in the right:

paraffins	benzene, xylene, and toluene
olefins	paraffins
aromatics	ortho-, para-, and meta-xylene
saturates	C_nH_{2n+2}
unsaturates	C_nH_{2n-6}
isomers	C_nH_{2n}
cyclics	butylenes

2. How many isomers of pentane, C_5H_{12} are there? Draw them.

3. An Ethyl Group is:

   a. A selection of three types of leaded gasoline

   b. $-CH_3$

   c. $-C_2H_5$

   d. $-C_2H_6$

   e. four little old ladies with the same name

4. If you've got a headache by now, you might take some acetyl salicylic acid. That's aspirin, and it has the chemical structure:

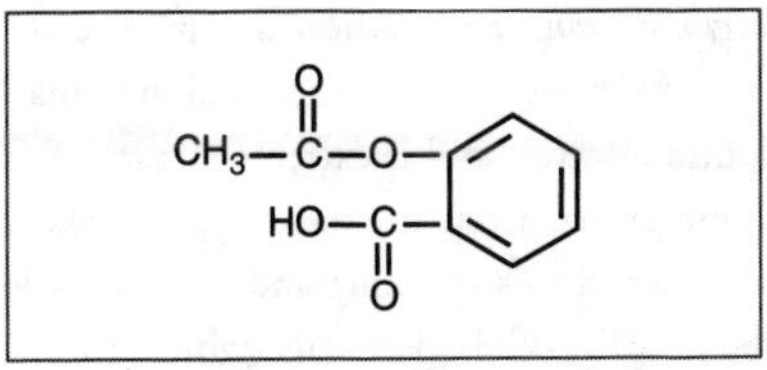

Fig. 1-17 Acetyl salicyclic acid

Find some examples of a methyl group, an ester group, a benzene ring, and an acid group in this molecule.

Endnotes for Chapter 1

[1]Often the terms *organic group* and *radical* are incorrectly used interchangeably. A radical looks like an organic group, except it can stand alone, unattached to a molecule. As a result, it has an unpaired, odd electron and is extremely reactive. The methyl radical, $\cdot CH_3$, can be produced, with some effort, from methane by the loss of one hydrogen atom. In writing, groups (which are attached to something and not stand-alone) are usually designated with a dash in front of them, as in $-CH_3$. Radicals (which are stand-alone) have the dot in front of them, as in $\cdot CH_3$. Also in writing chemical formulas, the shorthand use of the letter R (which is the first letter of the word radical) is common practice. The fact that R stands for (in this case) an organic group (not a stand-alone radical) doesn't help dispel the confusion between the two. (Even this explanation had to be rewritten in the third edition before it really sounded right.)

CHAPTER 2 ⬡

Benzene

"Oh my how many torments
lie in the circle of a . . . ring"

The Double Gallant,
Colley Cibber, 1671–1757

Why start out with benzene? The obvious answer is that benzene is one of the handful of basic building blocks in the petrochemicals industry, along with ethylene, propylene, and a few others. The more subtle reason is that benzene, more than any of those other chemicals, comes from a broader base— steel mill coking, petroleum refining, and olefins plants. For that reason, the benzene "network," the sources and the uses, is more complex than any of the others.

After a little historical background, this chapter will cover benzene production (including the hardware) as a chemical engineer might look at it, some of the important properties from the chemists' point of view, and the major benzene applications.

AN HISTORICAL PERSPECTIVE

Michael Faraday first isolated and identified benzene in 1825, during his scientific heydays at the Royal Institute in London. Benzene proved to be an enigma to chemists for more than a century after that. The valence rules of carbon and hydrogen require that benzene molecules have its characteristic alternating single and double bonds in the carbon ring. It baffled scientists that the benzene molecule didn't behave in the precise way that other molecules with double bonds did. In chemical reactions, the carbon-to-carbon bonds in the benzene ring acted in some ways like an average of single- and double-bonded carbons.

In 1865 the German scientist August Kekule offered a very appealing theory. He suggested the single and double bonds continuously trade places with each other—they oscillate or resonate. In the early 1930s, the famous Linus Pauling offered more convincing evidence supporting Kekule's theory, using quantum mechanics. There are still some loose ends, but no good alternate theory has turned up yet.

Benzene had limited commercial value during the 19th century. It was used primarily as a solvent. In the 20th century, gasoline blenders discovered that benzene had good octane characteristics. As a consequence, there was a large economic incentive to recover all the benzene that was produced as a by-product of the coke ovens in the steel industry. Starting around World War II, chemical uses for benzene emerged, primarily in the manufacture of explosives. Not only did the coke-oven benzene get diverted from gasoline blending to the chemical industry, but by midcentury, the refining industry itself was diverting huge quantities of benzene from gasoline blending stocks to keep up with chemical needs. Ironically, the largest consumer of benzene, the petroleum industry, ultimately turned out to be the largest supplier.

The increasing demands for benzene by the petrochemicals industry led to new and improved manufacturing processes—catalytic reforming, toluene dealkylation, and the newer toluene disproportionation, the last two being techniques for converting toluene to benzene. Toluene dealkylation has gone in and out of vogue as the economic winds have blown to and fro. A fortuitous source emerged in the 1970s when olefin plants started using heavy gas oil as a feedstock and produced by-product benzene.

BENZENE FROM COAL

An important raw material used in the manufacture of steel is coke, a nearly pure form of carbon. To supply themselves with coke, steelmakers developed the process of destructive distillation of coal.

The chemical makeup of coal is predominantly a mixture of very high molecular weight, polynuclear aromatic compounds. That mouthful is a common expression used in describing heavy hydrocarbon compounds. High molecular weight refers to the number of atoms, in this case carbon and hydrogen, attached together in the molecule. Ethane, C_2H_6, would be low molecular weight; $C_{30}H_{30}$ would be high molecular weight. Polynuclear aromatic refers to the preponderance of C_6 type rings in the molecule, as you can see in Figure 2–1.

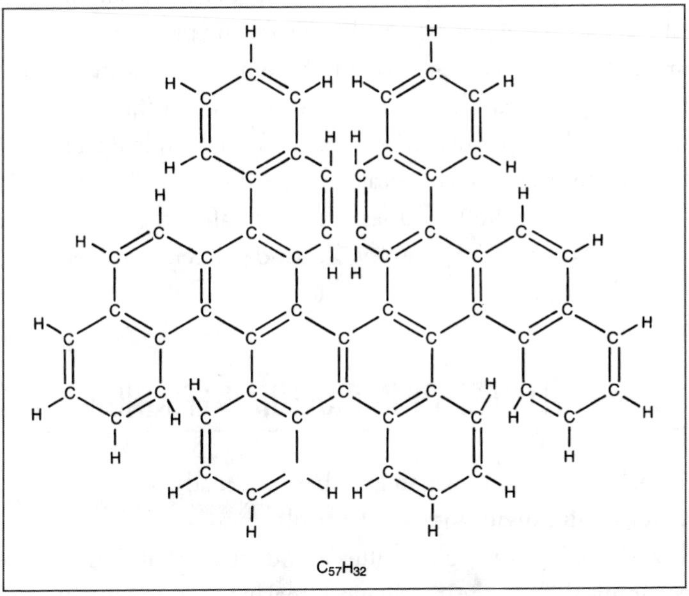

$C_{57}H_{32}$

Fig. 2–1 A polynuclear aromatic

Because of the size of these molecules and the multiple ring feature, the ratio of carbon to hydrogen is high, compared to other hydrocarbons encountered up to this point. In ethane it's 1:3; the compound in Figure 2–1 is almost

2:1. In the destructive distillation process, the coal is heated to 2300–2700°F in the absence of air. At those temperatures, the large molecules begin to crack, forming on the one hand, smaller organic compounds—many of which are liquids or gases at room temperature—and on the other hand, pure carbon, which is coke for the steel furnaces.

Because of the high carbon/hydrogen ratio, one ton of coal yields about 1500 pounds of coke and about 500 pounds of coal gas, coal oil, and coal tar. Prior to the advent of electricity, coal gas was a primary source of municipal lighting, and gaslights lined the streets of the great cities in 1900. Coal tar is a solid at room temperature and is often used as a roofing material or for road paving.

The coal oil is a mixture of benzene (63%), toluene (14%), and xylenes (7%), resulting directly from the benzene ring remaining intact during the cracking process. For this reason, steel companies became important suppliers of BTXs (benzene, toluene, and xylenes). None started out to get into the chemical business. They just exploited a valuable by-product.

Benzene from coal coking started to become less important in the 1950s as the benzene market mushroomed considerably faster than the steel market, and the marginal supply of benzene came from petroleum refining. Coal-based benzene for the U.S. chemical industry dropped from nearly 100% in 1955 to 50% in the 1960s and less than 5% after the 1980s. Coal-based economies like South Africa and New Zealand still rely considerably more on coal-derived benzene.

BENZENE IN PETROLEUM REFINING

The rapid increase in demand for benzene made obvious to oil refiners the advantages and disadvantages of petroleum as a supply source. Refiners were always looking for higher valued products; an all-liquids system was more economical than the mechanical coal processing system; and many chemical companies were subsidiaries of oil companies. At the same time, there was a limited amount of benzene naturally available in crude oil. It was the development of sophisticated refining processes for increasing yields of gasoline from crude oil that boosted benzene availability. The new processes created benzene out of other molecules, permitting them to be recovered along with the benzene naturally found in crude oil.

The benzene content of crude oil that comes out of the ground is typically only about 0.5–1.0%. Generally that's not enough to justify the equipment necessary to extract the benzene from the crude oil. Catalytic reforming became the more important and commercial source of benzene and by the end of the century accounted for about 50% of U.S. production. The original and still primary purpose of this process was to make high quality gasoline components out of low octane naphtha by reforming the molecules with the help of a catalyst. The feed to a catalytic reformer, naphtha, is a mixture of paraffins, naphthenes, and aromatics compounds in the C_6 to C_9 range. (*Naphthenes* is a predominantly refining word meaning saturated cyclics.) Typically a catalytic (cat) reformer changes the naphtha composition. In the process, as illustrated in Figure 2–2,

- paraffins are converted to iso-paraffins
- paraffins are converted to naphthenes
- naphthenes are converted to aromatics, including benzene

These are the good things that happen in the cat reforming process because iso-paraffins, naphthenes, and aromatics each have higher octane numbers than the molecules from which they were created. Other changes happen that are not so good.

- paraffins and naphthenes can crack to form butane and lighter gases
- some of the side chains (usually methyl groups) groups attached to the naphthenes and aromatics can break off also to form butanes and lighter gases

The results of both are lower octane and less valuable compounds than before.

The typical change in the composition of naphtha as it passes through the reformer is shown in Table 2–1.

	Volume Percent	
	Feed	Product
Paraffins	50	35
Naphthenes	40	10
Aromatics	10	55

Table 2–1 Composition change in a cat reformer

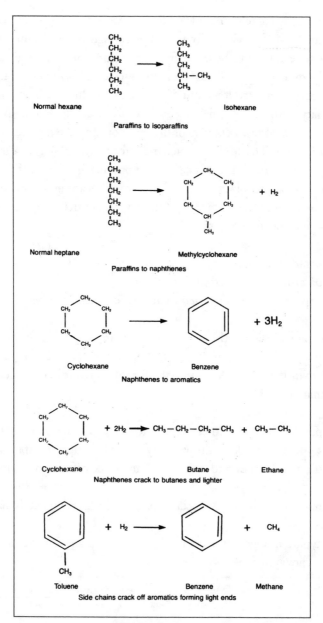

Fig. 2–2 Cat reforming reactions

The process

The reforming reactions are all promoted and controlled in the reactors illustrated in Figure 2–3. The naphtha is pumped through continuously at high temperatures (850–950°F) and pressures (200–800 psi), but even these severe conditions won't do it. The key ingredient is the presence of a catalyst. Each reactor is packed with pellets made of alumina or silica and coated with platinum, the catalyst. As the naphtha comes in contact with the platinum, various reactions take place, depending on the temperature and pressure in that particular reactor. Generally, there are several reactors, so different sets of operating conditions can be handled, each one aimed at promoting one of the desirable reactions listed above. The platinum catalyst, by the way, does not take part in the chemical reactions. It just promotes them. Old-timers in the industry like to chuckle and tell you that catalysts are a lot like some 10-year-old kids you know. They never get into trouble. It's just that wherever they go, trouble happens.

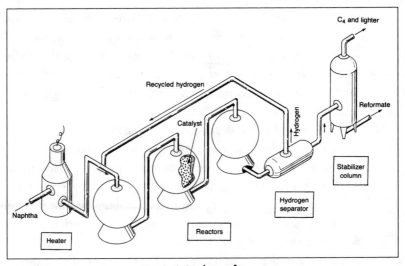

Fig. 2–3 Catalytic reformer

From the chemical equations in Figures 2–2, you can see that some of the reactions give off hydrogen, while others use it up. For this reason, when the product comes out of the last stage, hydrogen is separated, recycled, and remixed with the incoming feed. This provides an abundant supply during the reactions, which is necessary to prevent the formation of all but small amounts of coke during the cracking reactions. The coke will deposit itself on the catalyst, causing it to deactivate. The presence of excess hydrogen causes most of the coke to unite with the hydrogen to form light paraffins (mostly methane and ethane).

Eventually the catalyst becomes deactivated from coke buildup and miscellaneous other junk depositing on the catalyst, and the reactor must be shut down and regenerated. Otherwise the amount of conversion of feed to the desired products declines rapidly. Regeneration is done primarily by pumping very hot air through the reactor. The oxygen in the air reacts with the carbon on the catalyst forming carbon dioxide, which is then just blown into the atmosphere. Eventually, after a lot of deactivation and regeneration, the catalyst starts to collapse or become contaminated with other elements and must be replaced. Spent catalyst still contains all the original platinum, so it has a very high salvage value.

The amount of benzene produced in a reformer will depend on the composition of the feed. Every crude oil has naphtha with different PNA (paraffin, naphthene, aromatics) content. In commercial naphtha trading, the PNA content is often an important specification. High naphthene and aromatic content would indicate a good reformer feed. High paraffin content would indicate a good olefin plant feed.

The benzene yield also will depend on the mode in which the reformer is run. For example, setting the operating conditions to maximize benzene production will generally mean a sharp increase in the production of light ends—butanes and lighter gases. That's okay if you're not concerned about the loss of the other components, the ones used for gasoline. But if you are trying to maximize gasoline volume, benzene outturn may suffer.

The yields from a reformer, then, are a function of the feed composition and the operating conditions that are in turn responsive to economic incentives.

Downstream of the reactors and the hydrogen separator, the product is fed to one or more fractionating columns, where it is split into several streams. If just the butanes and lighter gases are removed, the remaining stream is generally called reformate. But in those refineries where benzene is recovered, to make subsequent processing easier, a "heart cut" that has all the benzene concentrated in a narrow boiling range, is removed from the reformate.

Often it is called, reasonably enough, benzene concentrate or aromatics concentrate. Benzene concentrate is about 50% benzene, plus some other C_5's, C_6's, and C_7's. All of them boil at about 176°F, the boiling point of benzene. Since the boiling temperature of the benzene is so close to that of the other hydrocarbons in the concentrate stream, simple fractionation is not a very effective way of isolating the benzene from benzene concentrate. Instead, one of two processes is used to remove the benzene, solvent extraction process or extractive distillation. The two differ in the primary mechanism they use. One operates on a liquid-liquid basis, the other on a vapor-liquid basis.

Solvent extraction

There are certain compounds that have the remarkable characteristic of being able to selectively dissolve some compounds, while at the same time ignoring others. A familiar example might be to take a spoonful of table salt and drop it into a half a glass of paint thinner. The salt sinks to the bottom of the glass. Mix it, shake it, and it still settles down to the bottom because it won't dissolve in paint thinner. Having observed that, it would be tough to get that salt completely separated from the paint thinner.

Now add a half a glass of water and stir. The salt disappears as it dissolves in the water. Now all you have to do is to separate the paint thinner and water by carefully pouring off the paint thinner, which has floated on the top. Then you just need to let the water stand for a couple of days and evaporate. In the bottom of the glass you've got nearly all the salt you started with.

In petrochemical language, in this example:

—salt-laden paint thinner is a concentrate

—water is the solvent

—salt is the extract

Solvent extraction of benzene works the same way. But instead of water, the various solvents used are sulfolane, liquid SO_2, diethylene glycol, and NMP (N-methyl pyrrolidone). The paint thinner/salt/water process described above might be called a batch solvent process, since it consists of sequential steps that can be repeated, batch after batch. Some low-volume commercial processes still operate that way.

In the analogy for benzene, a batch of benzene concentrate is mixed with the solvent; the benzene dissolves in the solvent; the solvent separates naturally from the undissolved components; the benzene-laden solvent is then drawn off and fractionated to separate the benzene. (This step is designed to be easy by selecting a solvent that has a boiling temperature much different from the benzene.) The fractionation products are solvent and benzene.

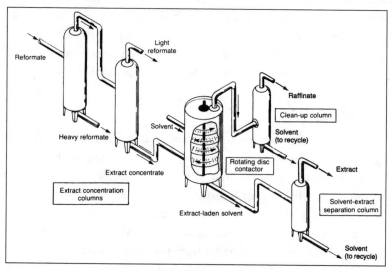

Fig. 2–4 Solvent extraction

Knowing how the batch process works, you'll find the continuous flow process just as simple. Figure 2–4 is a generic solvent extraction unit. In this case the extract is benzene. To make the process efficient, the benzene is concentrated by fractionating the reformate stream to the narrow boiling range around benzene. The benzene concentrate is pumped into the bottom of a vessel with a labyrinth of mixers inside. Sometimes the mixers are mechanically moved to achieve better extraction effectiveness. A rotating disc contactor is illustrated in Figure 2–4. The solvent is pumped in the top. Almost all the heavier solvent works its way to the bottom; the lighter benzene concentrate works its way to the top. As the two slosh past each other, the benzene is extracted from the concentrate, dissolving into the solvent.

The benzene-laden solvent is handled just like the batch process—it is fractionated to separate the benzene from the solvent; the solvent is recycled back to the mixing vessel.

The remnant hydrocarbons that are taken from the top of the mixing vessel are often called benzene raffinate, a misleading, ironic name. Benzene raffinate contains no benzene. It's the leftovers after the goodies are removed, but it is still a good gasoline blending component.

Extractive distillation

In the incessant scramble to reduce capital and operating costs, chemical engineers adapted a related technique for removing benzene from benzene concentrate. For years, absorption, a gas/liquid extraction process, has been used for separations in refinery gas plants and natural gas plants. It only took a technique for using the special absorbents, the same ones used in solvent extraction, to reduce the complexity of the equipment and the processing costs. (*See* Figure 2–5)

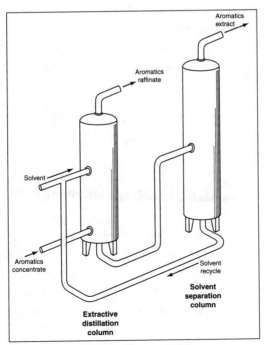

Fig. 2–5 Extractive distillation

Like the solvent extraction process, extractive distillation relies on the intimate contact of the liquid solvent and the aromatics concentrate vapors to allow the aromatics to be preferentially dissolved in the solvent. The usual list of solvents includes DEG (Diethylene glycol), TEG (Triethylene glycol), NMP (N-methyl pyrrolidone), or methyl formamide.

Again, as in solvent extraction, it's more efficient to concentrate the aromatics by fractionating a benzene concentrate, sometimes called a "heart cut," from the cat reformate. Then in Figure 2–5, the aromatics concentrate is heated and introduced as a vapor to the bottom section of the extraction column. The solvent of choice enters at the top of the column. The boiling point of any of the solvents used is high enough that it will remain liquid even as it trickles past the rising vapors of the aromatics concentrate. The trays inside the column are designed to cause that intimate contact necessary for the solvent to extract the benzene (and other aromatics, if the aromatics concentrate is cut that way) from the concentrate.

The aromatics-laden solvent leaves the bottom of the column as a liquid. The vapors leaving the top of the column, the aromatics raffinate, have almost no aromatics left. They are cooled and condensed to a liquid and used elsewhere, normally as a gasoline blending component.

The aromatics-laden or "fat" solvent is fractionated in a distillation column. The widely different boiling points of the solvent and aromatics make the separation relatively easy and clean. The solvent is recycled back to the beginning of the process. The aromatic extract, called crude benzene, is usually passed through a clay treater to remove any olefins that sometimes get created in the process and then distilled once again to produce high purity benzene.

Benzene from Olefin Plants

In Chapter 4 you'll find a complete discussion of the manufacture of ethylene and propylene by cracking naphtha or gas oil in an olefin plant. One of the by-products of cracking those feedstocks is benzene. The term "by-product" may not be appropriate anymore, since about a third of the benzene supply in the United States now comes from olefins plants.

Naphthas and gas oils consist of molecules with carbon counts of 5 to 20 or more. The olefins are created by heating the molecules to a temperature where they crack, forming among other things the desired ethylene (C_2H_2) and propylene (C_3H_6). The larger carbon count molecules, C_{10} and higher, often contain multiple benzene rings, not too unlike the coal configuration described above. When the molecules break up, the benzene rings can be freed intact, forming benzene and other aromatics. The process is similar to the destructive distillation of coal, when it comes to benzene.

The benzene leaves the olefins plant fractionator mixed with the other gasoline components so it is handled the same way as a refinery stream. An aromatics concentrate is made and run through one of the two separation processes you just read about, solvent extraction or extractive distillation.

BENZENE FROM TOLUENE HYDRODEALKYLATION

Since toluene is nothing more than benzene with a methyl group attached, creating one from another is relatively easy. Benzene, toluene, and for that matter, xylenes too, are coproduced in the processes just described—coke making, cat reforming, and olefin plants operations. The ratio of benzene to the other aromatics production is rarely equal to the chemical feedstock requirements for the three. One method for balancing supply and demand is toluene hydrodealkylation (HDA). This process accounts for 10–15% of the supply of benzene in the United States and is a good example of what can be done when one or more coproducts are produced in proportions out of balance with the marketplace.

The word *hydrodealkylation* is less ominous that it appears. Alkanes are a synonym for paraffins; alkylation is the process of adding a paraffin group (like a methyl or ethyl group) to another compound. Dealkylation is nothing more than removing it. *Hydro-* indicates the replacement atom is hydrogen.

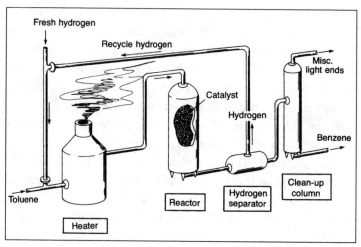

Fig. 2–6 Benzene from toluene hydrodealkylation

In the toluene HDA process shown in Figure 2–6, toluene is mixed with a hydrogen stream, heated, then pumped into a reactor. This vessel, like a cat reformer reactor, is packed with a platinum catalyst and runs at high pressures and temperatures. The methyl group pops right off as the toluene passes over the catalyst. Hydrogen fills out the valence requirements of the resulting molecule, forming benzene.

The stream leaving the reactor is separated in several fractionators into hydrogen, methane and other light gases, and benzene. The hydrogen is recycled, and the light gases are usually sent off to the fuel system. The benzene is usually clay-treated to remove any stray olefins and other contaminants, resulting in a pure, nitration grade benzene.

The yield of benzene in a toluene HDA plant runs 96–98%.

Material Balance

Feed:

Toluene	1200 lbs.
Hydrogen	27 lbs.

Product:

Benzene	1000 lbs.
Methane and other misc.	227 lbs.

BENZENE FROM TOLUENE DISPROPORTIONATION

In the last 15 years, as the demand for benzene and xylenes started to pull away from the demand of toluene, engineers and chemists scratched their heads and came up with a commercial process to increase the two at the expense of the one. Toluene disproportionation takes in toluene and turns out benzene and xylenes.

The definition of disproportionation involves the definitions of two other terms, oxidation and reduction, so here goes.

Oxidation and reduction: The term o*xidation* originally meant a reaction in which oxygen combined with another substance. It has broadened to include *any* reaction in which there is a transfer of electrons between substances. Oxidation and reduction always occur together, with the oxidizing agent gaining electrons and the reducing agent giving them up. This sounds like an oxymoron, but sorry, that's the way it is.

Disproportionation: A chemical reaction in which a single compound serves as both an oxidizing and reducing agent. It converts to a more oxidized and a more reduced derivative.

In the case of toluene disproportionation, reduction to benzene occurs when a methyl group pops off (hydrodealkylation takes place) and oxidation to xylene occurs as that methyl group that popped off attaches itself to another toluene molecule (a transalkylation reaction.)

$$2C_6H_5CH_3 \rightarrow C_6H_6 + C_6H_4(CH_3)_2$$

The secret that makes this process work is no surprise, the catalyst. Those that work include some of the noble metals, specifically, platinum or palladium, a rare earth metal like cerium or neodynium (are they rare or what?) on alumina, or a non-noble metal like chromium on a silica-aluminum support.

In Figure 2–7, toluene is fed into a heated reactor containing the catalyst in a fixed bed. A small amount of hydrogen is pumped in to keep carbon deposition on the catalyst to a minumum. The reactor conditions are in the 650–950°C and 150–500 psi ranges. The effluent is cooled then the hydrogen is recovered and recycled. The rest of the effluent is then triple distilled, removing nonaromatics in the first step, benzene in the second, and xylenes in the third.

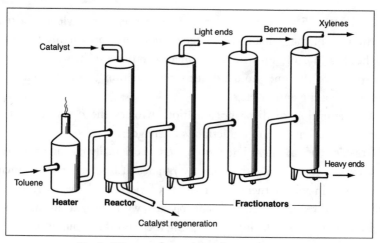

Fig. 2–7 Benzene from toluene disproportionation

The conversion rates on a once-through basis are high as you can see from the material balance, 2337 pounds of benzene/xylenes from 2400 pounds of toluene.

Material Balance

Feed:

Toluene	2400 lbs.
Hydrogen	7 lbs.

Product:

Benzene	1000 lbs.
Xylenes	1337 lbs.
Methane and other misc.	62 lbs.

HANDLING BENZENE

Benzene is a clear, colorless, flammable liquid with a distinct, sweet odor. It burns with a smoky flame, as do other hydrocarbons with high aromatic content. (That's why kerosenes with high aromatic content do not make good jet fuel or burning grade kerosene—too much black smoke.) Benzene is only slightly soluble in water.

Benzene Properties	
Molecular weight	78.11
Freezing point	41.9°F (5.5°C)
Boiling point	176.2°F (80.1°C)
Specific gravity	0.879 (lighter than water)
Weight per gallon	7.32 lbs/gal

The commercially traded grades of benzene are motor or industrial, pure (2°F boiling range), and nitration grade (1°F boiling range). The boiling range is a measure of the amount of impurities (other hydrocarbons) mixed in with the benzene. The wider the boiling range, the more impurities. Industrial pure benzene has about 0.5%; nitration grade has even less because it is clay-filtered to remove the more reactive compounds like thiophene, a sulfur-containing, bad-smelling heterocyclic. Motor benzene is generally mixed into gasoline, so it can stand the boiling range of 7°F.

Benzene is shipped in tank cars, tank trucks, barges, and drums. Transfers from one vessel to another are in closed systems because benzene is a poisonous substance with acute toxic effects. It'll kill you in 5–10 minutes if you breathe too much. Red DOT flammable liquid labels are required.

USE PATTERNS

Most of the benzene used in chemical applications ends up in the manufacturing processes for styrene (covered in Chapter 8), cumene (covered in Chapter 7), and cyclohexane (covered in Chapter 4). Polymers and all sorts of plastics are produced from styrene. Cumene is the precursor to phenol, which ultimately ends up in resins and adhesives, mostly for gluing plywood together. The production of styrene and phenol account for about 70% of the benzene produced. Cyclohexane, used to make Nylon 6 and Nylon 66, is the next biggest application of benzene.

Other smaller but important volumes of benzene end up in the processes for making maleic anhydride (for resins), nitrobenzene (for explosives), aniline (for dyes), and dodecylbenzene (for detergents).

Chapter 2 in a nutshell ...

Benzene, C_6H_6, is a ring of carbon atoms connected alternately by a single and double bond. Each carbon has a single hydrogen attached. It is found as a natural component in crude oil; it is created in the process of catalytically reforming naphtha to make high octane gasoline components; and it is formed in thermal cracking processes such as an olefins plant where complex molecules containing benzene rings are split up. Benzene is also made by hydrodealkylation of toluene and by disproportionation of toluene. High purity benzene is produced by either a solvent extraction process or extractive distillation. Benzene is used in the production of numerous chemicals including styrene, cumene, cyclohexane, and maleic anhydride.

EXERCISES

1. Assemble the following list into a table of feeds, operating units, and outturns:

benzene	coal	naphtha
benzene	coke	olefin plant
benzene	gas oil	toluene
benzene	destructive distillation	solvent extraction unit
benzene/xylene	toluene disproportionation	toluene
cat reformer	hydrodealkylation	reformate

2. If you had 500,000 gallons of toluene and a toluene HDA unit and the toluene market price was $0.20/lb., benzene was $0.24/lb., hydrogen was $0.40/lb., and it cost $0.005/lb. to run the HDA unit, what would you do? Oh, and toluene weighs 7.21 lbs. per gallon, and benzene is 7.32 lbs/gallon.

3. What is the "cat" in cat reformer?

4. What's the difference between reformate and raffinate?

5. Some coffee companies use methylene chloride to take the caffeine out of regular coffee (and you drink that stuff?). In this solvent extraction process, what do you think are the solvent, the raffinate, the extract, and the feed?

CHAPTER 3 ⬡

Toluene and the Xylenes

"Into fire, into ice."

Divine Comedy
Dante, 1265–1321

Should this be a separate chapter? The chemistry and hardware involved in making toluene and xylenes are for the most part the same as their sibling, benzene. While that may be true, there are a few chemical principles that can be demonstrated better using toluene. The separation processes for purifying toluene and xylenes are different also. There's enough, then, for a healthy bite without tagging on to the last chapter.

TOLUENE

The manufacture routes to toluene, like benzene, include cat reforming, olefin plant production, recovery of the small amounts naturally occurring in crude oil, and coke production. More than two-thirds of toluene comes from cat reforming. The volume of coal-derived toluene, which evolves in the same manner that was described in the benzene chapter, almost rounds off to zero now.

In the cat reforming process, two important variables control toluene make: the composition of the feed and the operating conditions in the reactors. As to the first, some compounds are more suitable for reforming into toluene than others. These precursors (from the Latin *curro*, I run, and *pre*, before) include cyclohexane, methyl cyclohexane, ethyl cyclopentane, and dimethyl cyclopentane.

In Figure 3–1, you will notice that three of these compounds have the same carbon/hydrogen count, C_7H_{14}, and the same carbon count as toluene, C_7H_8. Three different types of reactions take place in a cat reformer that change the precursors to toluene: ring opening, dehydrogenation, and cyclicization. Just looking at Figure 3–1, you can imagine that dehydrogenation (the removal of hydrogen) is necessary to work on the methyl cyclohexane. Because ethyl cyclopentane and dimethyl cyclopentane start out with the wrong carbon number in their rings, both ring opening and cyclization (closing a ring back up again), as well as dehydrogenation, are needed to get to toluene.

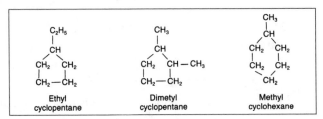

Fig. 3–1 Toluene precursors

When the naphtha feed to a cat reformer has a naturally high content of these precursors, the yields of toluene are high. Other than this fortuitous circumstance, there's generally not too much attention paid to toluene in the reforming operation for several reasons:

a.   more toluene generally means less benzene
b.   the composition of the naphtha feed depends on the selection of the crude oil, and that is usually determined by factors other than reformer operations because the reformer feedstock fraction is only a small part of crude oil
c.   most reformate ends up as a gasoline blending components (One of the components that makes it attractive is toluene, and most toluene is left in the reformate. Making more toluene isn't the issue—extracting it from the reformate is, and that is a matter of economic trade-off between gasoline and chemical uses.)

Separation of toluene from the other components can be by solvent extraction or extractive distillation, just as described in the benzene chapter. The boiling points of benzene and toluene are far enough apart that the feed to separation unit of choice can be split (fractionated) rather easily into benzene concentrate and a toluene concentrate. Alternatively, the separation unit can be thought of as aromatics recovery unit. Then an aromatics concentrate stream is fed to the solvent extraction unit, and the aromatics outturn can be split into benzene and toluene streams by fractionation. Both schemes are popular.

Azeotropic distillation of toluene

There is an alternate process for recovering toluene from the reformate stream called azeotropic distillation. It also can be used to split toluene from the other hydrocarbons that have boiling points near toluene. Azeotropic distillation is like solvent extraction with an extra twist. The process can be more efficient than extraction when the toluene concentration is high.

In this process, a solvent is used that increases the volatility of the components to be removed. In this case, what is removed is everything in the toluene concentrate but the toluene. The added solvent, along with the unwanted components, goes up the distilling column as a vapor; the toluene goes down and out as a liquid.

An analogy might help. Water in the gas tank of your car can cause a problem. It's usually caused by warm moist air getting in a half-empty gas tank, followed by the water condensing when the weather turns cold. Or maybe you just left the gas cap off when you ran through the car wash. In any event, water in gasoline causes hard starts and sputtering because it won't vaporize easily. Dry gas is the over-the-counter remedy, nothing more than ethyl alcohol. Water will dissolve in the alcohol, and together they will act just like gasoline as they go through the engine because together they vaporize at a lower temperature than either water or ethyl alcohol alone. In this analogy, the solvent is ethyl alcohol, the (toluene) extract is gasoline, and the raffinate is water.

When azeotropic distillation is used for toluene, the solvent used is usually a mixture of methyl ethyl ketone (MEK) and water (10%). The solvent and the toluene are mixed, heated, and then charged to a distillation column (*see* Figure 3–2). The paraffins and naphthenes dissolve in the MEK/water and then vaporize about 20°F lower than their normal boiling temperature. The vapors work their way up the distilling column; the toluene works its way down as a liquid. Again, this takes place despite the fact that the paraffins and the naphthenes have nearly the same boiling temperatures as toluene. The solvent does it.

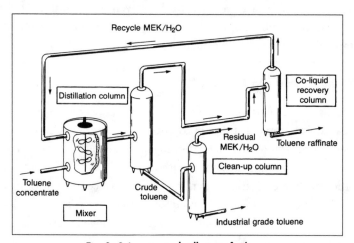

Fig. 3–2 Azeotropic distillation of toluene

Commercial use

Toluene, like benzene, is a flammable liquid and requires the red DOT shipping label. There are two commercially available grades, usually defined in terms of boiling ranges. Industrial grade toluene (95 to 98%) boils within two degrees of the toluene boiling point of 213°F (100.6°C). Nitration grade toluene (99%) boils within a one-degree range. The term "nitration" grade is a hangover from the specification required for the manufacture of trinitrotoluene (TNT). Lower grades are known as technical grade toluene, crude toluene, aviation grade toluene, or other specialty names.

Use patterns of toluene

During World War II, two militarily but not chemically related uses gave a running start to toluene. Because of its high-octane characteristics

(103–106 octane number), toluene was particularly suitable for blending aviation gasoline. Wartime conditions made maximum production of toluene an imperative during this period.

At the same time, the need to manufacture military explosives created a demand for toluene to make TNT. Ironically, the chemistry that makes for good octane characteristics has little to do with that of explosives.

In the postwar period, the expansion of commercial aviation sustained the demand growth for toluene as a high octane-blending component. By the 1960s, aviation gasoline gave way to kerosene-based jet fuel in most commercial aircraft. But the growth of automotive gasoline and the accompanying octane wars more than compensated, and today gasoline remains a major application for toluene.

To the dismay of toluene lovers, if there are any, the volume growth of benzene has overshadowed that of toluene, and toluene's major use is to make benzene in hydrodealkylation and toluene disproportionation units. About 50% of the toluene recovered in the United States is used this way. Conversion to para-xylene is also of growing importance.

Toluene is used more commonly than the other BTXs as a commercial solvent. There are scores of solvent applications, though environmental constraints and health concerns diminish the enthusiasm for these uses. Toluene also is used to make toluene diisocyanate, the precursor to polyurethane foams. Other derivatives include phenol, benzyl alcohol, and benzoic acid. Research continues on ways to use toluene in applications that now require benzene. The hope is that the dealkylation-to-benzene or disproportionation steps can be eliminated. Processes for manufacturing styrene and terephthalic acid—the precursor to polyester fiber—are good, commercial prospects.

XYLENE

The manufacture of the xylenes is a *déjà vu* story of benzene and toluene—cat reforming, olefin plants, a small amount naturally resident in crude oil, and coke making. A small but rapidly growing amount of xylene comes from catalytic disproportionation, the process described in the benzene chapter in which the methyl group is clipped off one toluene molecule (forming benzene) and ends up on another (forming xylene.)

What really makes the xylenes different from the other BTXs are the techniques to separate them from each other. That will be the main topic addressed in this section.

The BTXs from the refining and petrochemical industries follow more or less the same processing scheme (*see* Figure 3–3). In a refinery, the reformate stream coming from the cat reformer contains most of the aromatics, together with miscellaneous naphthenes (alicyclics) and paraffins (aliphatics). The BTXs in the stream coming from an olefins plant also has some olefins mixed in as well. The BTXs can be recovered as mixture or separately in a solvent extraction unit or an extractive distillation unit. (For those of you who read this section first and out of order, you are excused and may read the details in the benzene chapter.) Once the aromatics are separated from the rest, they need to be separated from each other.

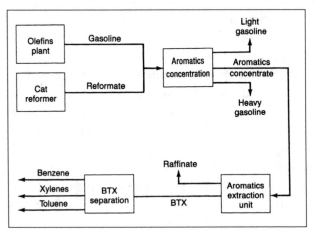

Fig. 3–3 Aromatic processing

The other X

You might think you've finished your introductions to the aromatics family, having dealt with benzene, xylene, and the xylene triplets, ortho-, para-, and meta-xylene. There's another isomer in the closet, ethylbenzene.

It has the xylene carbon/hydrogen count, C_8H_{10}, but it's a benzene ring with one ethyl group ($-C_2H_5$) attached, not two methyl groups ($-CH_3$). Moderate amounts of by-product ethylbenzene get created in the various chemical routes to the BTXs.

	Catalytic Reforming	Olefins Plant	Disproportionation
Ethylbenzene	26%	52%	-
Para-xylene	14%	10%	26%
Meta-xylene	41%	25%	50%
Ortho-xylene	19%	19%	24%

"On-purpose" ethylbenzene manufacture will figure importantly in the chapter on styrene.

The boiling points shown here of benzene, toluene, ethylbenzene, and ortho-xylene are different enough that they can be split apart without too much effort by fractional distillation in columns similar to those shown in Figure 3–4. "Too much effort" means extensive refluxing, reboiling, and energy consumption. Meta- and para-xylene can't because their boiling points differ by less than two degrees.

	Boiling Points °F	Freezing Points °F
Benzene	176.2	41.9
Toluene	231.4	-138.9
Ortho-xylene	292.0	-13.0
Meta-xylene	282.4	-54.2
Para-xylene	281.0	55.9
Ethylbenzene	277.1	-138.0

Ortho-xylene can be separated by distillation; ethylbenzene is only 3.9°F from para-xylene, but by using very tall, multitrayed distillation columns (200 feet high with 300 trays), it too can be separated fairly completely. The 1.4°F spread between meta- and para-xylene requires columns more expensive than chemical companies can stand, so alternate separation techniques have been developed: cryogenic crystallization and adsorption using molecular sieves and isomerization.

Fig. 3–4 Haltermann Custom Processing Plant at Houston, Texas

CRYOGENIC CRYSTALLIZATION

Even though the boiling temperatures of meta- and para-xylene are close together, their freezing points, i.e., the temperatures at which the liquid starts freezing, i.e., turning to crystals, are not. Meta-xylene crystallizes at -54.2°F and para-xylene at +55.9°F, a spread of more than 100 degrees.

In Figure 3–5, the processing scheme shows the ortho-xylene and ethylbenzene split out in fractionators. The mixed para- and meta-xylenes are then processed in a fashion a lot like making good pot roast gravy. In order to keep the grease out of the gravy (and if you've got the time), you can put the beef drippings in the refrigerator for an hour or two. All the grease solidifies and floats to the top and can be spooned off and discarded. Similarly, the mixed para- and meta-xylenes are cooled initially to about

-90°F in a holding tank. At that temperature, para-xylene crystals form and grow in a liquid-solid mixture like slush. The key to good solid-liquid separation is large crystal growth. The larger the crystals, the better the separation because of the next step.

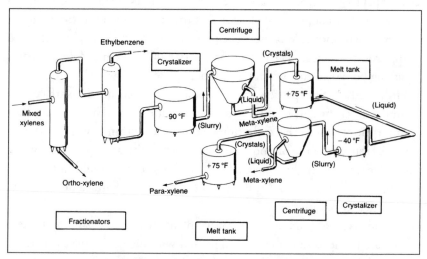

Fig. 3–5 Xylene separation and cryogenic crystallization

When the crystals have grown sufficiently, the slush is put in a centrifuge. The spinning action permits the para-xylene to separate from the mother liquor, so-called because the crystals come out of the liquid. At this stage, the para-xylene crystals, called *filter cake* at this point, have a purity of 80-90%, due to the mother liquor that coats the crystal surface. (That's the reason for big crystals—less surface area for the mother liquor to coat.)

To further purify the para-xylene, the crystals are again melted, cooled—this time to about -40°F—and crystallized once again. Centrifuging this time results in about 99% pure para-xylene. The meta-xylene from both centrifuges is about 85%, the rest para-xylene.

ABSORPTION

A commercial process using a material called molecular sieve can also separate para-xylene from meta-xylene. Molecular sieves are marble-sized pellets that have millions of pores, all of a size that para-xylene molecule can fit in but the meta-xylene molecule cannot. The pore sizes are so small they are measured in Angstroms, which are $1x10^{-8}$ centimeters (0.00000001 cm). Molecular sieves of varying pore sizes are used in many other applications as well.

In an adsorption process, molecules are collected on the surface of a substance or agent and held there by electrostatic force. That is in contrast to absorption, where the collection takes place within the agent, like a solvent. Hence, the uncommon suffix *ad*. When the adsorption agent is a molecular sieve, it might seem that the collection of the liquid in the pores is more like *ab-* than *ad-*. But the collection is on the surface of the pores, so don't get too confused.

In this type of para-xylene recovery plant, the mixed para- and meta-xylenes are pumped into a vessel bed packed with molecular sieve pellets. At first, the liquid stream coming out of the bed (the effluent stream) is very low in para-xylenes, because they are selectively being collected in the sieves. Gradually the concentration of para-xylene in the effluent stream starts to build as the molecular sieve pores fill up and some para-xylene slips by. At some point, the operation is shut down. A fluid is then pumped backwards through the bed to flush out the para-xylene. The process of pulling the para-xylene from the molecular sieve, *desorption,* uses a fluid called a desorbent to flush the para-xylene out. The desorbent used has a boiling point different enough from para-xylene that its separation from the para-xylene in a distillation column is easy.

After desorption, the bed is heated up to vaporize and remove all the desorbent and remaining para-xylene. The cycle is then ready to begin again. Para-xylene purity from this technique is about 99.5%. The adsorption process sometimes has an economic advantage over the cryogenic crystallization route, due mainly to fuel and operating costs.

DISPROPORTIONATION AND TRANSALKYLATION

In the chapter on benzene and in Figure 2–7, you saw that toluene disproportionation yielded both benzene and mixed xylenes. When the catalyst-prompted methyl group removes itself from the toluene, it usually attaches itself to another toluene molecule in a way that it forms xylene. That's transalkylation. The freed methyl group might attach itself momentarily to another free benzene molecule, or it might attach itself to the methyl group of another toluene, forming ethylbenzene. However, the creation of benzene and xylenes predominates, and the combined yields of the two are 92-97%.

$$2C_6H_5CH_3 \rightarrow C_6H_6 + C_6H_4(CH_3)_2$$

ISOMERIZATION

With the percentages of the three xylenes from the various sources differing so much, it's not likely that a company, or the industry for that matter, will produce just the amount of the xylene isomer it wants. Para-xylene has the biggest demand and meta- the smallest, but none of the processes, cat reforming, olefins plants, or disproportionation, have commensurate yields.

The research gnomes at the lab benches have now developed the catalysts, and companies have commercialized plants to shift a mixture of mixed xylenes and ethylbenzene towards the para- isomer and away from the meta-.

Licensors offer a variety of catalysts to promote the isomerization—silica alumina by itself or enhanced with a noble metal like platinum or a non-noble metal like chromium. Another uses hydrofluoric acid with boron trifluoride

In the case of the noble metal catalytic process, the feed enters a vessel with a fixed catalyst bed at 850°F and 14.5 psi. As is often the case, a small amount of hydrogen is present to reduce the amount of coke laying down on the catalyst. The effluent is processed in a standard fashion to separate the hydrogen, the para- and ortho-xylene, and any unreacted or miscellaneous compounds. Yields of para-xylene are in the 70% range.

Aromatic	Feed composition	Product composition
Ethylbenzene	26%	-
Para-xylene	14%	71%
Meta-xylene	41%	-
Ortho-xylene	19%	20%

Commercial use

Mixed xylenes are commercially available in nitration grades that have tolerances of 3 and 10°F, depending on the specified amount of the hydrocarbon present. Purities of the ortho-, meta-, and para-xylenes are more often than not a matter of negotiation between buyer and seller.

Mixed xylenes are used as an octane improver in gasoline and for commercial solvents, particularly in industrial cleaning operations. By far, most of the commercial activity is with the individual isomers. Para-xylene, the most important, is principally used in the manufacture of terephthalic acid and dimethyl terephthalate en route to polyester plastics and fibers (Dacron, films such as Mylar, and fabricated products such as PET plastic bottles). Ortho-xylene is used to make phthalic anhydride, which in turn is used to make polyester, alkyd resins, and PVC plasticizers. Meta-xylene is used to a limited extent to make isophthalic acid, a monomer used in making thermally stable polyimide, polyester, and alkyd resins.

The xylenes are flammable and are shipped under the same regulations and using the same methods as benzene and toluene: tank cars, trucks, barges, and tankers. Pipeline movements are limited. Toxicological problems dictate handling in closed systems like benzene and toluene.

Chapter 3 in a nutshell ...

Toluene, $C_6H_5CH_3$, and the xylenes, $C_6H_4(CH_3)_2$, are benzene rings with one or two methyl groups, -CH_3, hung on in the place of hydrogens.

Toluene has one; xylenes have two. Sources of all three BTXs are the same: crude oil, catalytic reforming, heavy liquids cracking in on olefins plant, and, to a declining extent, coking at a steel plant.

Most of the toluene and xylenes have their origin in catalytic reforming or olefins plants. From there, the processing schemes vary widely from site to site. The schematic in Figure 3–6 captures most of the variations, although it's hard to portray that some plants separate the BTXs from each other early in the scheme while others do it at varying places downstream of an aromatics recovery unit.

All the BTXs are high-octane gasoline blending components. In the petrochemicals business, toluene is used as a building block for polyurethane. Para-xylene and ortho-xylene are used to make polyester fibers and plastics, alkyd resins, and plasticizers.

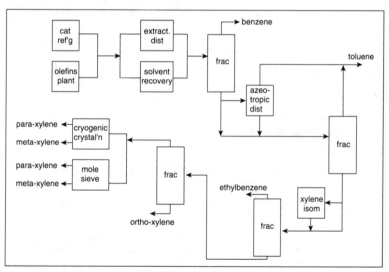

Fig. 3–6 Processing schemes for BTXs

EXERCISES

1. Name the six compounds that make up the BTXs.

2. Unmix and match the feeds and products for the following plants?

Feeds	Plants	Product
toluene	cryogenic distillation	mixed xylenes
toluene	cryogenic distillation	mixed xylenes
mixed-xylenes	disproportionation	ortho-xylene
mixed-xylenes	hydrodealkylation	para-xylene

3. Can a mixture of benzene, toluene, and meta-xylene be separated by cryogenic crystallization? What's the usual (more economic) way?

4. Ortho-, para-, and meta-xylene are (choose one):

   a. isomers
   b. aromatics
   c. solvents
   d. petrochemical feedstocks
   e. all of the above
   f. the three ugly daughters of a mad Hungarian named Vladok Xylene.

CHAPTER 4 ⬡

Cyclohexane

"A hen is an egg's way
of making another egg."

Life and Habit
Samuel Butler, 1835–1902

The petrochemicals business is funny. Some companies use cyclohexane to make benzene. Some use benzene to make cyclohexane. This chapter covers the latter.

The development of nylon by DuPont in 1938 generated the initial big commercial interest in cyclohexane as they settled on its use as their preferred raw material. In the period right after World War II, the manufacture of nylon grew for a while at 100% annually, quickly overwhelming the availability of cyclohexane naturally available in crude oil. The typical crude oil processed in U.S. refineries at the time had less than 1% content of cyclohexane. Ironically, since cyclohexane leaves the crude oil distillation operation in the naphtha, it was usually fed to a cat reformer, where it was converted to benzene. As it turned out, with so many other precursors also being converted to benzene in the cat reformer, benzene became a good source for cyclohexane manufacture.

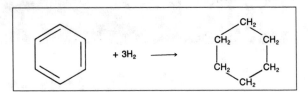

Fig. 4–1 Benzene hydrogenation to cyclohexane

As a source of cyclohexane, benzene has the right shape (*see* Figure 4–1) but too few hydrogens. So cyclohexane plants are not much more than vessels in which benzene molecules are hydrogenated with the help of a catalyst. This process accounts for about 90% of today's cyclohexane.

THE TRADITIONAL PROCESS

Benzene, you'll recall, has alternating double bonds, and the addition of one hydrogen atom to any one of the carbons will cascade quickly all around the benzene ring so that all the carbons pick up hydrogen. Pressure and temperature alone cannot cause the hydrogenation—a catalyst is needed. Fortunately, several metals qualify—platinum, palladium, nickel, and chromium. The first two are highly active and can cause hydrogenation to occur at room temperature and only 15 to 20 psi pressure. Unfortunately, platinum and palladium are expensive metals, and most commercial processes use nickel or chromium. Though they require much higher temperature and pressures (and more expensive in terms of energy costs), the catalyst is cheaper.

Sulfur and carbon monoxide can be killers (literally) with hydrogenation catalysts. It will "poison" them, making them completely ineffective. Some sulfur often shows up in the benzene feed, carbon monoxide in the hydrogen feed. The alternatives to protect the catalyst are either to pretreat the feed and/or the hydrogen or to use a sulfur resistant catalyst metal like tin, titanium, or molybdenum. The economic trade-offs are additional processing facilities and operating costs vs. catalyst expense, activity, and replacement frequency. The downtime consequences of catalyst replacement usually warrant the more expensive treatment facilities.

Refineries and olefins plants generate the primary supplies of benzene so cyclohexane plants tend to be clustered around refining centers to save transportation costs.

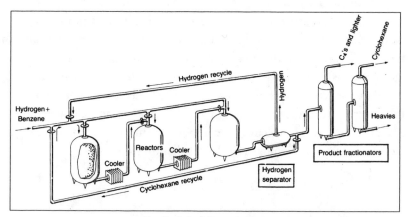

Fig. 4–2 Cyclohexane plant

The hardware used for the hydrogenation of benzene is shown in Figure 4–2. The basic parts are three or four reactors in a series plus a separation section at the end. The reactors are vessels filled with catalysts in the form of charcoal or alumina pellets that are coated with one of the above metals. The catalyst is packed loosely enough that the feed can flow through, top to bottom, by gravity.

The continuous flow process shown in Figure 4–2 has a mixture of benzene, cyclohexane, and hydrogen being heated to about 400°F, pressured to about 400 psi, and pumped through the first reactor. The proportions of each feed depend on the type of catalyst being used. On a once-through basis, about 95% of the benzene is converted to cyclohexane.

Most hydrogenation reactions, this one included, are exothermic, i.e., they give off heat. To minimize the by-products that could occur, strict temperature control must be maintained. As the feed passes through the reactor like the one shown in Figure 4–3, the temperature increases by about 50°F. The reactor effluent is therefore cooled back down to 400°F in a heat exchanger. For the second pass, additional benzene is added, although the resulting proportion in the second and succeeding reactors keeps decreasing. The same process of hydrogenation, with its exothermic effects, occurs and the reactor effluent must be cooled again in a heat exchanger to get it to the right temperature for the next reactor.

The overall conversion of benzene to cyclohexane is nearly 100%, but the effluent from the last reactor will still have plenty of hydrogen in it. To facilitate the hydrogenation reaction, hydrogen is usually kept in excess. The effluent is passed through a flash drum, where the pressure drops and the hydrogen flashes out of the product and is recycled to the feed. The remaining effluent is then fractionated as a final cyclohexane purification step. (Since reaction conditions in the process are never controlled perfectly, some of the benzene feed and whatever other hydrocarbons come along with it, get converted to other miscellaneous compounds, mostly butanes and lighter gases, that have to be removed in the distillation step.)

Fig. 4–3 Reactors at Conoco's cyclohexane plant at Lake Charles, Louisiana

A cyclohexane stream is recycled to the feed and also performs an important function. It acts as a heat sink or a sponge, diluting the exothermic effect of the hydrogenation reaction, keeping the temperature down. At temperatures about 450°F, the decomposition of benzene to those light ends just mentioned increases rapidly.

Material Balance

Feed:

Benzene	944 lbs.
Hydrogen	65 lbs.
Catalyst	–

Product:

Cyclohexane	1000 lbs.
Light Ends	9 lbs.

In summary, the key variables in this process are temperature control, excess hydrogen, and catalyst activity. Conversions are typically 99.5%.

THE LIQUID PHASE PROCESS

A cost savings–induced process captures much of the new plant design. The efficiencies come from running a liquid phase system, saving heat and catalyst handling equipment. Figure 4–4 shows the flows.

The reaction vessel is filled with a finely divided Raney nickel catalyst suspended in cyclohexane. (Raney nickel is a dark gray, pyrophoric—ignites easily and burns hot—powder named after the man who devised the process for extracting it from nickel-aluminum alloy.) The catalyst slurry is kept in suspension throughout the reaction by pumping it vigorously and circulating it through an external heat exchanger.

Benzene and hydrogen in excess of what's needed are fed cold into the pressurized, catalyst-filled reactor. The hydrogenation reaction that takes place, attaching six hydrogen atoms each to the benzene molecules, is exothermic.

The catalyst circulation through an external heat exchanger is set to keep the reactants at more or less a constant temperature of 350–400° F.

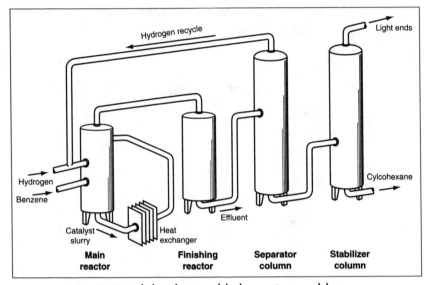

Fig. 4—4 Liquid phase benzene dehydrogenation to cyclohexane

A finishing reactor with a fixed bed of catalyst completes the catalytic hydrogenation of any residual, unreacted benzene. The effluent from this reactor is then cooled and flashed to remove most of the hydrogen and then fractionated to produce high purity cyclohexane.

One pass through this process at 400 psi results in 100% conversion of the benzene to cyclohexane with purity of about 99%. The economies compared to the traditional processing scheme come from energy savings and simple equipment. In addition, the catalyst circulation system lends itself to fine control since deactivated catalysts can easily be replaced on the fly without shutting down the system.

Phenol has for a long time been a minor source of cyclohexane, more so in Europe than in the United States. Phenol, a benzene ring with an -OH group attached in place of a hydrogen, is a coproduct of the manufacture of acetone. Ironically, the process starts with benzene, as you can read about in Chapter 7. Only when the demands for acetone and phenol get out of sync and too much phenol is left over after the market clears itself does the phenol route to cyclohexane become an attractive proposition.

COMMERCIAL ASPECTS

Cyclohexane is a colorless, water-insoluble, noncorrosive liquid having a really pungent odor. It's flammable like any naphtha product, and it is shipped in tank cars, tank trucks, barges, and drums. Red DOT shipping labels are required. In commerce, trade is usually done on the basis of Technical Grade (either 95 or 99% purity) or Solvent Grade (85% minimum purity).

Cyclohexane Properties	
Molecular weight	84.16
Freezing point	43.7°F (6.5°C)
Boiling point	177.3°F (80.7°C)
Specific gravity	0.7786 (lighter than water)
Weight per gallon	6.54 lbs/gallon

Nearly all cyclohexane is used to make three intermediate chemicals. About 85% goes for caprolactam and adipic acid. Another 10% goes for hexamethylene diamine (HMD). All three are the starting materials for Nylon 6 or Nylon 66 synthetic fibers and resins. Nylon fiber markets include the familiar applications: hosiery, upholstery, carpet, and tire cord. Nylon resins are engineering plastics and are largely used to manufacture gears, washers, and similar applications where economy, strength, and a surface with minimum friction are important.

Cyclohexane is the starting point for making the chemical intermediates cyclohexanol and cyclohexanone. Other minor uses include industrial solvent applications such as cutting fats, oils, and rubber. Cyclohexane also makes a good paint remover component.

Chapter 4 in a nutshell ...

Cyclohexane, C_6H_{12}, is a carbon ring with two hydrogen atoms attached to each carbon. It resembles benzene, but there are no double bonds. Benzene is the feed to a cyclohexane plant, which is just a hydrogenation process.

Cyclohexane is a colorless liquid at room temperature. It is used primarily to make precursors of Nylon 6 and Nylon 66.

EXERCISES

1. What does "exothermic effect of hydrogenation reaction" mean?

2. Fill in the blanks:

   a. To make cyclohexane out of benzene, you need to add _____ atoms.

   b. Various _____ are used to promote the benzene hydrogenation reaction.

   c. If it weren't for the development of _____ most cyclohexane would still end up as a gasoline-blending component or as cat reformer feed.

3. If cyclohexane is worth 30 cents/lb., hydrogen costs 40 cents/lb., and it costs 0.5 cents/lb. of feed to run a traditional cyclohexane plant, how much can you afford to pay for benzene to break even? Assume the light ends are flared (that is, they are burned off and worth nothing).

4. Which came first, the chicken (cyclohexane) or the egg (benzene)?

CHAPTER 5 ⬡

Olefin Plants, Ethylene, and Propylene

"Anything that can happen
will happen."

Murphy's Third Law

OLEFIN PLANTS

The big daddy of the petrochemicals industry is the olefin plant. The vintage of the technology that dominates the scene dates back to the 1930s. Olefin plants are a wellspring of the industry's basic building blocks —ethylene, propylene, butylenes, butadiene, and benzene. The proportions of olefin plants built in the last few decades are huge. The so-called world-scale plant (the size that achieves whatever is currently considered full economies of scale) is larger than many medium size refineries. Capacity is no longer measured in millions but billions of pounds per year.

Olefins plants, for the most part, all have the same basic technology, but the process flows differ with the varied feedstocks that can be used. This chapter will cover in some depth the feeds, the hardware, the reactions, and the variables that can be manipulated to change the amount and mix of products. The physical properties of ethylene and propylene, which present some unique handling problems, will be covered also.

A handful of plants with completely different technologies have been built and are described in a section at the end of this chapter. They contribute only a minor amount of olefins to the marketplace, and their economics and outlooks are still under the microscope, so most of this chapter will concentrate on the popular design.

Traditional olefin plants have more than one alias. One is even fraudulent. They are variously called *ethylene plants* after their primary product; *steam crackers* because the feed is usually mixed with steam before it is cracked; or "*whatever*" *cracker*, where "*whatever*" is the name of the feed (ethane cracker, gas oil cracker, etc.). Olefin plants are sometimes referred to as ethylene crackers, but only those who don't know any better use that misnomer. Ethylene is not cracked but rather is the product of cracking.

The one aspect of ethylene manufacture that sets it apart from most other petrochemicals is the wide range of alternate feedstocks that can be used. Most others are limited to one or a few commercial alternatives. Ethylene is a simple molecule, $CH_2=CH_2$, and if you thought that lots of hydrocarbons could be cracked to form it, you'd be right, as you can see in Table 5–1.

Ethane and propane produce a high yield of ethylene. Propane also gives a high yield of propylene. The earliest commercial olefin plants of any size were designed to use these two feeds, and they dominated U.S. plant designs in much of the 20th century.

Because of the shortages of U.S. natural gas in the 1970s, a perception developed that future production of natural gas would be declining. Along with it, the availability of NGLs, the natural gas liquids (ethane, propane, butanes, and natural gasoline) that typically makes up 10–20 percent of natural gas, would be tighter. Propane from refineries would be expensive since it would be a substitute for the disappearing natural gas.

Furthermore, in the United States, the naphthas were needed for the growing gasoline market. For this reason, by the late 1970s, industry developed the technology and built the plants to crack gas oils. In the United States, about half the ethylene comes from ethane/propane cracking; most of the rest comes from naphtha and gas oil. In Europe and the Pacific Basin, where large deposits of natural gas and the natural gas liquids in it are not readily accessible and gasoline is a smaller part of the energy mix, naphtha has been the feedstock of choice. The huge natural gas fields in the Middle East support ethane cracking there.

Pounds Per Pound of Feed

	Ethane	Propane	Butane	Naphtha	Gas Oil
Yield:					
Ethylene	0.80	0.40	0.36	0.23	0.18
Propylene	0.03	0.18	0.20	0.13	0.14
Butylene	0.02	0.02	0.05	0.15	0.06
Butadiene	0.01	0.01	0.03	0.04	0.04
Fuel gas	0.13	0.38	0.31	0.26	0.18
Gasoline	0.01	0.01	0.06	0.18	0.18
Pitch	—	—	—	—	0.10

Table 5–1 Olefin plant yields

The process

Ethane and propane cracking is simpler than heavy liquid (gas oil or naphtha) cracking, and you need to tackle it first. When ethane is heated up to 1700°F or above, either of two basic reactions can occur, splitting of carbon-hydrogen bonds and splitting of carbon-carbon bonds. There is a popular adage in ethylene plant lore: "Anything that can happen will happen." The product from ethane cracking depends on which bond is cleaved, as shown in Figure 5–1. Ethylene forms from carbon-hydrogen fractures; methane forms from carbon-carbon cleavage with the resulting methyl radical picking up a hydrogen. Even acetylene and hydrogen might form and survive. But if the olefin plant is run right, the predominant yield will be ethylene.

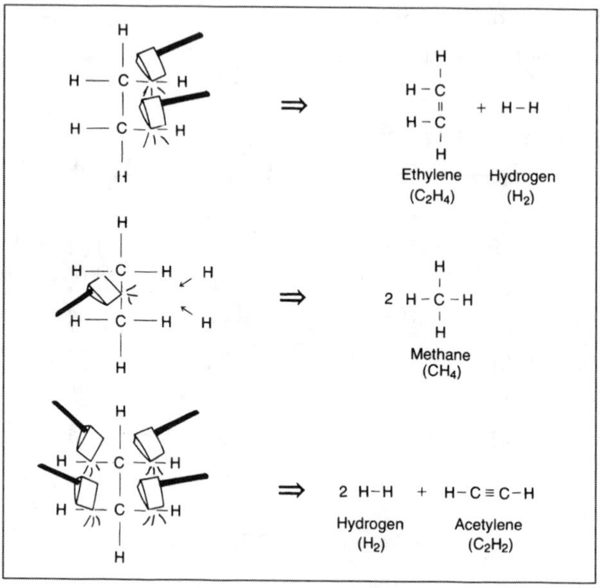

Fig. 5–1 Ethane cracking

Propane cracking is a little more complicated because there are more combinations and permutations of possible fractures. Not only is there ethylene and methane in the outturn, but also propylene and, surprisingly, ethane. Cracking propane at the carbon-carbon bond will give you ethane and methane (after the methyl and ethyl radicals pick up hydrogens) or ethylene if both go at the same time. Ethane also gets formed as a second round draft choice when two methyl radicals find each other instead of hydrogen. "Anything that can happen . . . "

When naphtha or gas oil is cracked, imagine the limitless combinations possible. Naphthas are made up of molecules in the C_5 to C_{10} range; gas oils from C_{10} to perhaps C_{30} or C_{40}. The structures include everything from simple paraffins (aliphatics) to complex polynuclear aromatics, so a much wider range of possible molecules can form. Ethylene yields from cracking naphtha or gas oil are much smaller than those from ethane or propane, as you can see from Table 5–1. But to compensate the plant operator, a full range of other hydrocarbons is produced as by-products also.

In the C_4 and C_5 range, a new breed of molecules shows up. These are aliphatics with more than one set of double bonds. The commercially important ones are butadiene and isoprene, both of which are important raw materials for making synthetic rubber. Butadiene is big enough to be covered in much more detail in the next chapter.

The BTXs, covered in the last two chapters, result mostly from cracking away the miscellaneous chains on complicated aromatic ring–containing molecules in the feed. Other olefin plant products in the C_5/C_6 and heavier category are mostly used as refinery process or blending stocks. Many of them have higher value than the naphtha or gas oil feed. So even though they are by-products, they contribute significantly to paying for the operation of the olefin plant. In fact, in many companies, heavy liquids crackers are the link that integrates the refinery with the petrochemical plant as one complex, with various streams going back and forth between the two.

The hardware

Olefin plants all have two main parts: the pyrolysis or cracking section and the purification or distillation section. The ethane cracker in Figure 5–2 has a pyrolysis (from the Greek, pyros, fire) section that consists of a gas-fired furnace where the cracking takes place. The newer individual furnaces can each handle more than 400 million pounds per year of ethane feed.

Ethane is pumped through a maze of 4- to 6- inch diameter tubes where it is heated up to about 1500°F and cracks. The ethane, by the way, never comes in direct contact with the fire. Otherwise, it would ignite. It stays inside the tubes.

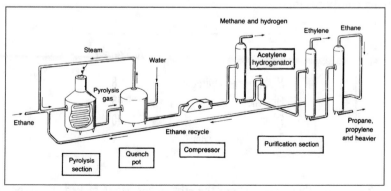

Fig. 5–2 Ethane cracker

The ethane is pumped through the pyrolysis section at a very high rate. Residence time of any individual molecules is a few seconds or less in the older plants and less than 0.1 second in the newer ones. This rapid rate is required to keep the cracking process from running away, resulting in the ethane cracking all the way to methane or even coke (carbon) and hydrogen. To further control this runaway cracking, the ethane is mixed with steam before it is fed to the furnaces.

Steam has two beneficial effects. First, it lowers the temperature necessary for the cracking to take place. That reduces the fuel bill and also the amount of methane and hydrogen that gets formed. Second, at the lower temperature, less coke forms and so less deposits on the inside of the furnace tubes. That saves having to shut down the furnaces for de-coking so often, a step necessary to prevent clogging and cold spots. Coke deposits act as an insulator, preventing the ethane from heating to the right temperature for effective cracking.

As soon as the hot effluent (the gases coming out of the reaction section) leaves the cracking furnace, it enters a quench pot. The gases coming out of the furnace are so hot they will continue to crack, just like a steak will continue to cook after you take it off the grill. So the gases are immediately hit with a stream of water to cool them down. The heat transfers from the gases to the water, which causes the water to turn into steam. This steam is subsequently recycled by separating it from the effluent and mixing it with the fresh feed to the furnaces.

At this point, the cracked gases consist of a mixture typically of the following composition:

	% Weight
Methane and Hydrogen	8
Ethylene	48
Ethane	40
Propane and heavier	4

You can see that only 60% of the ethane has been cracked. Forty percent of the effluent stream is still uncracked ethane. So part of the purification section will be dedicated to separating the ethane so it can be fed back to the furnaces again. This arrangement is sometimes referred to by the grizzly expression "recycling to extinction." So, while the pyrolysis section only makes 48% ethylene, recycling results in the combined pyrolysis/purifications yielding:

	% Weight
Methane and Hydrogen	13
Ethylene	80
Propane and heavier	7

(and no ethane!) The transformation is just arithmetic. Just take the one-through yields, drop the 40% ethane, and divide everything by 0.60 so they add up to 100%.

In the purification section of an ethane cracker, the gas can be handled in one of two ways. In order to fractionate the streams, they must be liquefied. Since they are all light gases, liquefaction can be done either by increasing the pressure in a compressor or by reducing the temperature to very low points in something called a "cold box." The ethane cracker in Figure 5–2 shows the compressor option. (Even then, the streams have to be cooled to assure they liquefy.)

Downstream of the compressor is a series of fractionators (generally the tallest towers in an ethylene plant) which separate the methane and hydrogen, the ethylene, the ethane, and the propane and heavier. All are heavy metallurgy to handle the pressures and insulated to maintain the low temperatures. There's also an acetylene hydrogenator or converter in there. Trace (very small) amounts of acetylene in ethylene can really clobber some of the ethylene derivative processes, particularly polyethylene manufacture. So the stream is treated with hydrogen over a catalyst to convert the little acetylene present into ethylene.

It may seem curious that an ethane cracker has propane and heavier included in the outturns. There are two reasons. The ethane used as feed is rarely pure. It generally has a couple percent of propane and heavier in it that results in a small amount of heavier products. But why bother or go to the expense to get pure ethane feed? In the first place, the olefins plant purification section can handle them. Secondly, some heavy hydrocarbons are actually formed anyway in the frantic scramble of free radicals and hydrogen that goes on during the cracking process.

Heavier feeds. As plant designers choose heavier feeds for the olefin plant, the hardware gets more extensive and expensive. The flow through the plant is still about the same as for the lighter feeds, as shown in Figure 5–3. In that simplified flow diagram, the heavy liquid feed goes to the pyrolysis section where it is cracked. Next, it goes to quench section where it is cooled, then to the separation section where it is split into its components.

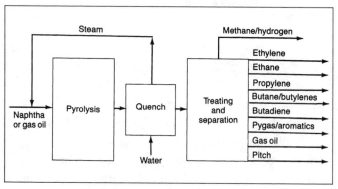

Fig. 5–3 Flows in a heavy liquid cracker

Referring to the hardware in Figure 5–4, there are much larger facilities required for heavier liquids cracking than for ethane or propane. As you saw in Table 5–1, the yield of ethylene from the heavier feeds is much lower than from ethane. That means that to produce the same amount of ethylene on a daily basis, the gas-oil furnaces have to handle nearly five times as much feed as ethane furnaces. As the design engineer scales up these volumes, he or she has to worry about the size of the tubes necessary to heat up that much feed, the residence times best for each kind of feed, and the best pressure/temperature/steam mixture conditions.

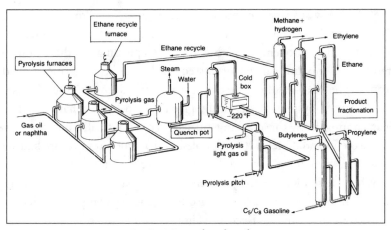

Fig. 5–4 Heavy liquid cracker

The separation section of a gas oil cracker looks like a small refinery, as you can see in Figure 5–4 or in Figure 5–5. In addition to the fractionators and treaters used in the purification section of the simpler ethane cracker, there are facilities to separate the heavier coproducts. In the front end of the separator facilities in Figure 5–4, the cold box option for handling the liquefaction of the gases is shown. Temperatures as low as -220°F are achieved in this super-refrigerator. At those low temperatures, Freon won't do the job. Liquid air, methane, ethylene, or ammonia are often used as the refrigerant in much the same way Freon has been used in an air conditioner.

Fig. 5-5 Olefins plant – heavy liquids cracker

Several new streams are introduced in Figure 5–4. Propylene handling will be covered in detail later in this chapter. The C_4 stream is a combination of butanes, butylenes, and butadiene. Depending on the commercial interest, this mixture can be processed further to separate the individual streams. The C_5^+ gasoline stream, usually called pygas (sounds like an ailment caused by pizza), is typically given a mild hydrogenation step. The pygas coming from the cracking furnaces contain small amounts of very reactive olefin and diolefin (two double bonds) structures that are bad actors in gasoline. They form gums and lacquers in car engines.

When these olefins are hydrogenated, the pygas stream with its high-octane number becomes a good gasoline-blending component. It can be processed in an Aromatics Recovery Unit to remove the BTXs, which can be up to 10% of the pygas stream. (*See* Chapter 2, Benzene From Olefin Plants.) But even after the aromatics are removed, it still makes good gasoline-blending stock.

The fuel oils coming out of olefin plants are also characterized by an abundance of polynuclear aromatic molecules. (Same definition as for Figure 2–1). They are sometimes inaccurately referred to as having a high aromatics content. Nomenclature aside, because of this, the burning characteristics of pyrolysis gas oil and pyrolysis pitch are poor. They are smoky, sooty, and gum formers; they tend to be more viscous, and because of their polynuclear aromatic content, they are suspected carcinogens. They are basically a witch's brew of unsavory hydrocarbons.

Usually when anything heavier than ethane—or sometimes heavier than propane—is cracked, there is a furnace designed to handle the ethane recycle stream. The plant shown in Figure 5–4 shows three heavy liquid furnaces and one ethane furnace. Since the alternate use for ethane is usually refinery fuel, the economics often dictate recovery and cracking.

Process variables

Despite the abundance of analysis in the technical journals on the subject, there really is only a moderate amount of flexibility to change the yields in olefin plants once they are built. The problem is that the yield of each of the coproducts moves in a different direction as the pressures and temperatures, and residence times in the furnaces are changed. The fluctuations of the economic values of the by-products often result in little incentive to effect yield changes.

More significant, however, are the changing values of the feedstocks. In many plants or companies, the design permits substituting one feed for another; say ethane for propane or naphtha for gas oil. In those cases, plant operations respond to the market for feedstocks and products and reflect themselves in the changing yields implied by Table 5–1.

Other technologies

Methanol dehydrogenation to ethylene and propylene. In some remote locations, transportation costs become very important. Moving ethane is almost out of the question. Hauling propane for feed or ethylene itself in pressurized or supercooled vessels is expensive. Moving naphtha or gas oil as feed requires that an expensive olefins plant with unwanted by-products be built. So what's a company to do if they need an olefins-based industry at a remote site? One solution that has been commercialized is the dehydrogenation of methanol to ethylene and propylene. While it may seem like paddling upstream, the transportation costs to get the feeds to the remote sites plus the capital costs of the plant make the economics of ethylene and its derivatives okay.

The methanol dehydrogenation process consists of three sections, a fluidized bed reactor, a catalyst regeneration column, and product recovery columns, as shown in Figure 5–6. A fluidized bed reactor is designed to let the catalyst travel along with the feed, promoting the reaction as it moves. In this case, methanol mixes with the catalyst that comes in the form of a powder so fine that it behaves like a fluid. By the time the methanol has passed through the reactor, 99% has converted to ethylene and propylene in a strange shift of carbon and hydrogen atoms.

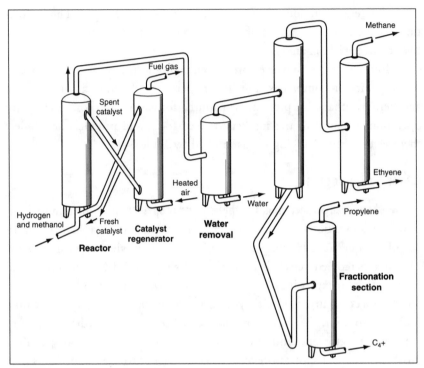

Fig. 5–6 Dehydrogenation of methanol to ethylene and propylene

The catalyst in this case is a silico-alumino-phosphate molecular sieve. Each particle has millions of pores, each the size of 4 Å or less. (Å or angstrom is equal to approximately 10^{-8} centimeters and is a measure used to gauge the size of single molecules. Tiny indeed.) The catalyst is buried in those 4 Å pores, so the size olefins that can get created are mostly ethylene and propylene. The catalyst is therefore highly selective in generating its products.

In the process, the catalyst is continuously separated from the mixture at the top of the reactor and sent to a regenerator where carbon and other contaminants are removed by blowing hot air into the regenerator forming CO/CO_2. From the top of the reactor, the product mix moves to the separation columns where ethylene and propylene are separated from the heavier C_4^+ products, any unreacted methanol, and the water that gets formed.

By varying the operating conditions, the pressure, and temperature, this process can produce a wide range of ethylene to propylene ratios, from a ratio of 0.75 to 1.5, as shown in Table 5–2.

Material Balance for Methanol Dehydrogenation

Feed:

Methanol	2000 lbs.	2000 lbs.

Products:

Ethylene	672 lbs.	912 lbs.
Propylene	890 lbs.	600 lbs.
Butylenes & heavier	366 lbs.	262 lbs.
CO_2 & coke	32 lbs.	86 lbs.
Other	40 lbs.	120 lbs.

Table 5–2

Propane dehydrogenation to propylene. If you were a company with lots of propylene applications, you might find that the traditional olefins plant is an unfocused way to satisfy your own business needs. Besides propylene, it makes coproducts galore, all of which have to be disposed of profitably. That makes your propylene derivatives businesses vulnerable to many economic variables. Finally, someone has figured a way to reduce that uncertainty by using the next closest thing to propylene—propane—as the feed and producing virtually all propylene by (just) removing two of the hydrogen atoms. That, of course, doesn't remove all economic uncertainty since propane is not without its own price volatility. It does, however, concentrate the mind on just two hydrocarbons: propane and propylene.

In this process, propane, and a small amount of hydrogen to control coking, are fed to either a fixed bed or moving bed reactor at 950–1300° F and near atmospheric pressure. Once again the catalyst, this time platinum on activated alumina impregnated with 20% chromium, promotes the reaction. In either design, the catalyst has to be regenerated continuously to maintain its activity.

The usual columns and stabilizers separate the reactor effluent. Unreacted propane and some hydrogen is recycled and mixed with the fresh feed. The net result is about 85% propylene, 4% hydrogen, with the rest light and heavy off-gases.

As the growth rates of propylene outpace ethylene, and companies restructure and simplify their portfolios, this catalytic route becomes more attractive.

Metathesis of ethylene and butylenes to propylene. Another on-purpose route to propylene is metathesis, a chemical reaction that starts with two compounds, involves the displacement of groups from each and produces two new compounds. The application in this case converts ethylene and mixed butylenes to propylene and butene-1. This route could appeal to a company with refinery or olefins plant ethylene and butylenes that both have market values less than propylene, which could be the case in some local markets.

The process calls for feeding ethylene and mixed butylenes to the bottom of a reactor. The mixed butylenes consist of both butene-1 and butene-2. (Refer to Figure 1–10 to refresh your memory about the difference.) A slurry of rhenium-based catalyst is introduced at the top. As the ethylene and butylenes bubble past the catalyst, the ethylene and butene-2 will react to form propylene (the carbon count is right). Simultaneously, as the butene-2 is consumed, butene-1 isomerizes to create more.

The reactor effluent is fractionated to produce a high purity propylene stream and recycle ethylene and butylene streams. Selectivity to propylene is greater than 98%. That is, 98% of the converted ethylene and butylene ends up as propylene so the process has few unwanted by-products.

ETHYLENE

Ethylene is a colorless gas with a slightly sweet odor. It turns from liquid to gas at -155°F. It burns readily in the presence of oxygen with a luminous flame. In fact, it was the ethylene component that made coal gas so useful as a gas light fuel at the turn of the 19th century. The other components in the coal gas don't give off near the light when burned by themselves. Natural gas lamps or propane/butane lanterns must be fitted with mantels to reduce the oxygen available, permitting only partial oxidation. That process gives off light. Burning ethylene doesn't require a mantel.

Ethylene Properties	
Molecular weight	28.05
Freezing point	-169.2 °C
Boiling point	-103.7 °C
Density at 0°C	0.95

The logistics of ethylene are tough. Because it's a light gas, high pressures or extremely low temperatures are required to handle it as a liquid. Very little ethylene is transported by truck and even that has to be done under special permit. For long hauls, the trucker will generally have to vent off some of the ethylene to keep the rest of it cool enough so that is can be contained at a reasonable pressure. Venting off ethylene (or any gas for that matter) is a cooling process. Ever wet your finger and stick it up in the air to see which way the wind was blowing? Whichever side of your finger got cool was where the wind was coming from. That's because the moisture on your finger was vaporizing, and that's a cooling process. In order for the moisture to go from liquid to vapor, it must pick up heat from its surroundings (your finger). Similarly, when ethylene is vented from a truck, it takes heat from the remaining ethylene. And were not the temperature kept down, the pressure of the ethylene would increase dangerously.

Ethylene trucks are expensive to build and operate. The fuel, the ethylene loss, and the energy necessary to liquefy the ethylene are costly factors. So most transport of ethylene is by pipeline. Although the operating costs of a pipeline are low, the initial construction costs are high. Pipeline transport, like most other aspects of ethylene, is capital intensive. For this reason, most consumers of ethylene are located in proximity to the producers of ethylene.

Ethylene pipelines operate more like natural gas pipelines than petroleum pipelines. That is, the ethylene moves as a gas, not as a liquid. The critical temperature is the reason. There's an interesting physical phenomenon involved here. Every gas has a critical temperature, and if you keep the gas above that temperature, no matter how much you increase the pressure, the gas won't liquefy. (The reason is a complex explanation having to do with atomic structure.) The critical temperature for ethylene is 48.6°F. Ethylene pipelines are usually buried about 10–15 feet below ground level, so the surrounding temperature is always 60–70°F. Ethylene, then, can be pumped around at very high pressures, 700–800 psi, and at those temperatures it is a very dense gas—nearly as dense as liquid, but still a gas.

Storage of ethylene is also an expensive proposition. For small volumes, like transfer tanks in a chemical plant, cylindrical or spherical tanks are often used. The pressure requirements at normal temperatures demand heavy duty, thick, expensive steel vessels. Storage of any size, say beyond 100,000 pounds, warrants cryogenic storage (from the Greek *kryos*, cold; and *gen*, bring forth). Cryogenic tanks are much lighter and cheaper steel tanks. Their use is made possible because the ethylene is supercooled way below the critical temperature (*see* Figure 5–7). Under this condition, the ethylene is liquid and very little pressure is needed to keep the ethylene from vaporizing. The operating cost of cryogenic tanks is high. Despite the fact that there is thick insulation around the tank, some heat leaks into the ethylene. To keep the ethylene below -155°F, the boiling point, some of the vapor is drawn off the tank, passed through a refrigeration unit where it's liquefied, then returned to the tank.

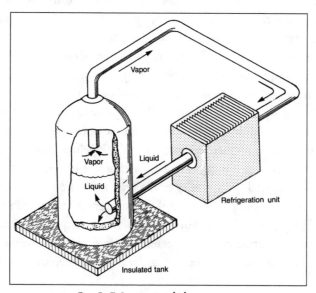

Fig. 5–7 Cryogenic ethylene storage

Circulating this stream faster or slower through the refrigeration unit keeps the liquid ethylene temperature in balance with the change in temperature outside the tank.

For large inventories of ethylene, in the millions of pounds, underground storage has been found very cost effective. It usually takes the form of caverns mined in rock, shale, or limestone or jugs leached out of salt in large underground salt domes as shown in Figure 5–8.

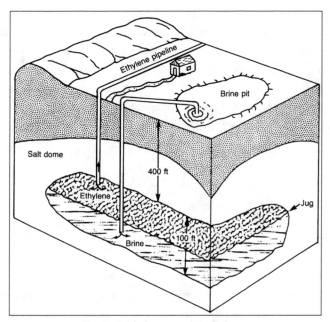

Fig. 5–8 Ethylene jug in a salt dome

In a jug, the more common facility, ethylene is moved in and out by displacement. When ethylene is pumped in, it displaces the brine (salt water) in the jug. To remove ethylene from the jug, brine is pumped in, displacing the $C_2^=$. Like any other hydrocarbon/water combination, the ethylene and water do not mix. So the water acts as a pressuring agent on the ethylene. Jugs or caverns are generally located a couple hundred feet below ground level. Ground temperature is a constant temperature, 65–70°F and is always above the critical temperature of ethylene. So the weight of the water in the standpipe is enough to keep the ethylene compressed and at a normal, not cryogenic, temperature for pipeline transport.

The cost of salt dome construction is cheaper than mined-cavern storage, which in turn is a lot cheaper, per pound of ethylene, than cryogenic storage and pressure storage.

The chemical uses for ethylene prior to World War II were limited, for the most part, to ethylene glycol and ethyl alcohol. After the war, the demand for styrene and polyethylene took off, stimulating ethylene production and olefin plant construction. Today's list of chemical applications for ethylene reads like the "What's What" of petrochemicals: polyethylene, ethylbenzene (a precursor to styrene), ethylene dichloride, vinyl chloride, ethylene oxide, ethylene glycol, ethyl alcohol, vinyl acetate, alpha olefins, and linear alcohols are some of the more commercial derivatives of ethylene. The consumer products derived from these chemicals are found everywhere, from soap to construction materials to plastic products to synthetic motor oils.

PROPYLENE

Propylene, like ethylene, is a colorless gas at room temperature. It is as flammable as LPG (liquefied petroleum gas or propane). In fact, propylene can be used as a substitute or supplement to LPG. The fuel characteristics are nearly indistinguishable. However, the petrochemicals industry bids propylene away from the fuels market and gives it a much higher price than LPG.

Propylene is traded commercially in three grades: refinery, chemical, and polymer grades. The difference is almost entirely the ratio of propylene to propane in the stream. Refinery grade propylene usually runs about 50–70% propylene; chemical grade 90–92%; polymer grade is at least 99% propylene. The remaining percentage is almost all propane in each case.

The reasons for the three grades are very practical. For the first two, refinery and chemical, that's the way they're made. Refinery grade propylene streams are generally by-products of a refinery's cat cracker, and the propane/propylene ration is determined by the way the cat cracker is run to make gasoline, not propylene. Chemical grade propylene is usually produced in a naphtha or gas oil cracker. The ratio of propylene and propane is about 92:8 over most of the operating conditions.

Some applications, particularly polypropylene manufacture, require very pure propylene feed. Polymer grade propylene is made by simple fractionation of one of the less pure propylene streams, refinery or chemical grade.

The logistics of propylene are more conventional than ethylene, but still expensive. While ethylene is like natural gas, propylene handling and logistics are almost identical to LPG. At room temperature, propylene has to be kept in a pressurized container to keep it from evaporating. It boils at -54°F, so cooling it down to keep it liquid is expensive.

Propylene Properties	
Molecular weight	42.08
Freezing point	-301.5°F (-185.3 °C)
Boiling point	-53.9°F (-47.7 °C)
Density at 0°C	0.51

Propylene is moved in equally large volumes by pipeline, tank car, and tank truck. All three modes handle propylene as a liquid, operating at pressures of about 200 psi. The storage facilities for large volumes of propylene are the same as those for ethylene, underground jugs, or caverns. Because of the lower pressure requirements than ethylene, cryogenic storage is rarely used. Storage in the form of steel spheres (typically 5–10 million pounds) and cylinder or "bullets" (200–500 thousand pounds) are prevalent.

Unlike ethylene, more propylene has always been produced than has been needed for the chemical industry. The situation goes back to the advent of thermal cracking units in refineries in the early part of the 20th century. By World War II, with catalytic cracking units generating larger volumes of by-product propylene, chemists had been challenged sufficiently to develop both petrochemical applications and refinery uses for propylene. As a consequence, there has since been a large amount of propylene used in the manufacture of gasoline. The most popular process has been alkylation, in which a high octane C_7 hydrocarbon is made by reacting propylene with isobutane in the presence of sulfuric or hydrofluoric acid. The product is called propylene alkylate and has an octane number of about 96, so it is a good gasoline blending component.

The propylene equivalent of polyethylene is polypropylene. About 50% of the chemical use of propylene is directed to that use. Other major applications are the manufacture of propylene oxide, isopropyl alcohol, cumene, oxo alcohols, acrylic acids, and acrylonitrile. The consumer products you are familiar with show up everywhere: carpets, rope, clothing, plastics in automobiles, appliances, toys, rubbing alcohol, paints, and epoxy glue.

Chapter 5 in a nutshell ...

Ethylene, C_2H_4, and propylene, C_3H_6, are both the smallest and the biggest petrochemicals. They are the largest volume petrochemicals; they have the simplest structure (at least ethylene does). Their most attractive feature is the double bond between two carbon atoms, which makes them highly chemically reactive.

Cracking large hydrocarbons usually results in olefins, molecules with double bonds. That's why the refinery cat crackers and thermal crackers are sources of ethylene and propylene. But the largest source is olefin plants where ethylene and propylene are the primary products of cracking one or more of the following: ethane, propane, butane, naphtha, or gas oil. The choice of feedstock depends both on the olefins plant design and the market price of the feeds.

In an olefins plant, the feed is subjected to very high temperatures in cracking furnaces for a few moments and then cooled rapidly to stop the cracking. Elaborate separation facilities are necessary to separate the olefins from the by-products of the cracking process.

Some new process technologies involving the use of catalysts to reduce costs are becoming popular, but traditional steam cracking dominates the olefins market.

Both ethylene and propylene are gases at room temperature and are handled in pressurized, closed systems. The list of derivatives of these two building blocks is impressive.

EXERCISES

1. To make at least 500 million pounds of ethylene per year and at least 200 million pounds of propylene per year, how much propane or gas oil would you have to crack in an olefins plant? How much butadiene would you make in either case?

2. Why doesn't the ethylene burn up as it passes through the furnace?

3. List all the possible products that could result from cracking butane, CH_3-CH_2-CH_2-CH_3, in an olefins plant. Remember, "Anything that can happen . . ."

4. Why do you suppose olefin plants are never referred to as propylene plants? Is propylene a second-class petrochemical?

5. Name four alternatives to the steam cracker to make ethylene and/or propylene.

CHAPTER 6 ⬡

The C₄ Hydrocarbons Family

"If you cannot get rid of the family skeleton,
you may as well make it dance."

George Bernard Shaw, 1856–1950

The first serious notice of C_4 hydrocarbons came with the development of refinery cracking processes. When catalytic cracking became popular, refiners were faced with disposing of a couple of thousand barrels per day of a stream containing butane, butylenes, and small amounts of butadiene. Their first thought was to burn it all as refinery fuel, but then they developed the alkylation process. With that, they could undo some of the molecule shatter that took place in the crackers and reassemble some of the smaller pieces as alkylate, a high-octane gasoline-blending component.

During World War II, the Japanese cut off U.S. access to sources of natural rubber, giving the Americans a strategic imperative to develop and expand the manufacture of synthetic rubber. The C_4 streams in refineries were a direct source of butadiene, the primary synthetic rubber feedstock. As a coincidence, the availability of this stream was growing rapidly with the expansion of catalytic cracking to meet wartime gasoline needs. Additional butadiene was manufactured by dehydrogenation of butane and butylene also.

In the 1950s, U.S. olefin plants that were cracking propane and butane began to produce modest amounts of by-product C_4 hydrocarbon streams. In Europe and Japan, and later in the U.S in the 1960s and 1970s, gas oil- and naphtha-based olefins plants rivaled refineries in the volume of the C_4 streams being produced.

A typical C_4 hydrocarbon stream coming from a gas oil or naphtha cracker, like that shown in the last chapter in Figure 5–4, might have the following composition:

Isobutane	5%
Normal butane	5%
Butadiene	42%
Isobutylene	18%
Butene-1	18%
Butene-2	12%

The terms butene-1 and butene-2 are the petrochemical nomenclature used to refer to what the petroleum refining industry calls normal butylenes. Butene-1 and -2 are more specific and descriptive. But to complicate matters more, there are two kinds of butene-2, cis-butene-2 and trans-butene-2.

In Figure 6–1, if you look closely, you'll see that the difference between butene-1 and the butene-2 is the location of the double bond. Butene-1 has it at the end position/ butene-2 at the middle. The methyl groups in *trans*-butene-2 are across from each other, on opposite sides of the fence; in the *cis* form they are next to each other or on the same side of the fence. The difference is more than cosmetic. It determines the way the molecule behaves, physically and chemically. Check the boiling points, for instance, in Figure 6–1. They differ, and that helps in the separation process. In a few paragraphs, the different applications for butylenes that derive from their chemical behavior differences will be covered.

Configuration	Name	Boiling Temperature, °F
$CH_3 - \overset{\displaystyle CH_3}{\underset{\displaystyle |}{CH}} - CH_3$	Isobutane	10.9
$CH_3 - \overset{\displaystyle CH_3}{\underset{\displaystyle |}{C}} = CH_2$	Isobutylene	19.6
$CH_3 - CH_2 - CH = CH_2$	Butene-1	20.7
$CH_3 - CH = C = CH_2$	1, 2 butadiene	51.4
$CH_2 = CH - CH = CH_2$	1, 3 butadiene	24.1
$CH_3 - CH_2 - CH_2 - CH_3$	Normal butane	31.1
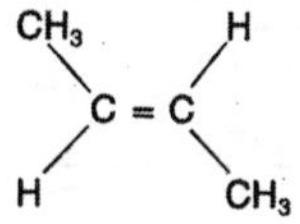 $\underset{H}{\overset{CH_3}{>}} C = C \underset{CH_3}{\overset{H}{<}}$	Trans-butene-2	33.6
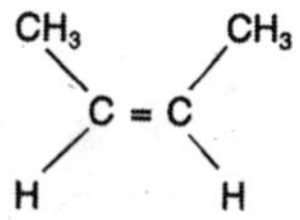 $\underset{H}{\overset{CH_3}{>}} C = C \underset{H}{\overset{CH_3}{<}}$	Cis-butene-2	38.7

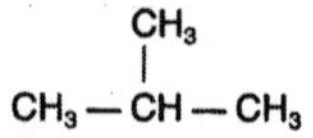

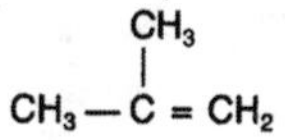

Fig. 6–1 Characteristics of the C4 Hydrocarbons

The structural difference between the two butadienes is pretty obvious in Figure 6–1, but most mixtures of butadiene are predominantly 1,3 butadiene, and there is little attention paid to the difference between the two.

PROCESSING

There are a dozen different ways to handle the C_4 stream in a petrochemical plant if you follow all the combinations possible in Figure 6–2. Simple fractionation won't do it because the boiling temperatures are so close together. Generally the first step is to remove the butadienes by extractive distillation, of the kind shown in Chapter 3.

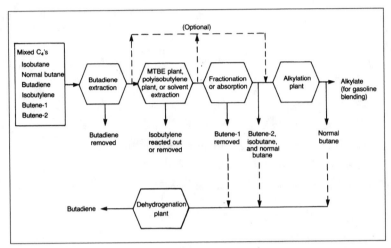

Fig. 6–2 Typical C$_4$ processing scheme

Isobutylene is the most chemically reactive of the butylene isomers. If the objective is just to get the isobutylene out of the C$_4$ stream, it can be removed by reaction with methanol (CH$_3$OH) to make MTBE (methyl tertiary butyl ether), by reaction with water to make TBA (tertiary butyl alcohol), by polymerization, or by solvent extraction. After that, butene-1 can be removed by selective adsorption or by distillation. That leaves the butene-2 components, together with iso- and normal butane, which are generally used as feed to an alkylation plant.

Not all chemical plants have all these facilities. Furthermore, some plants have processes to convert butanes to butadienes; others convert them the other way. The best way to sort out the options is to treat them one at a time.

BUTADIENE

Butadiene is one of the group of big four petrochemical building blocks, in company with ethylene, propylene, and benzene. It is used primarily as a feedstock for synthetic rubber, elastomers, and fibers. Butadiene is a colorless gas at room temperature but is normally handled under pressure or refrigerated as a liquid.

The base-load supply of butadiene is from olefins plants simply because butadiene is coproduced with the other olefins. There's not much decision on whether or not to produce it. It just comes out, but in a small ratio compared to ethylene and propylene. Cracking ethane yields one pound of butadiene for every 45 pounds of ethylene; cracking the heavy liquids, naphtha or gas oil, produces one pound of butadiene for every seven pounds of ethylene. Because of the increase in heavy liquids cracking, about 75% of the butadiene produced in the United States is coproduced in olefin plants.

As chemical companies rely more heavily on ethane and propane feeds to their olefins plants to generate their ethylene and propylene supplies, the coproduction of butadiene in olefins plants has not kept up with demand. Industry has resorted to building plants that make "on-purpose" or swing supply butadiene. The processes involve catalytically dehydrogenating (removing hydrogen from) butane or butylene.

Dehydrogenation

The one-step process for making butadiene from butane or butylene involves passing the feed over a catalyst at about 1200°F and under reduced pressure. The catalysts used are ferric oxide, aluminum-chromium oxide, or calcium-nickel phosphate. Depending on the feed, two or four hydrogen atoms pop off the butane (C_4H_{10}) or butylene (C_4H_8) molecules, forming butadiene (C_4H_6).

$$C_4H_{10} \rightarrow CH_2=CH-CH=CH_2 + 2H_2$$

$$C_4H_8 \rightarrow CH_2=CH-CH=CH_2 + H_2$$

The butane and/or butylene feed exposure to the high temperatures has to be controlled to a short interval, about 0.25 seconds. The effluent from the reactor must be quenched immediately to prevent the butadiene from continuing to "cook" and cracking to smaller light ends.

After the quench, the stream is compressed, and the butadiene is recovered through fractionation. Typical yields in this process are 60–65%.

A variation on this process improves the yields to as much as 89%. The process of cracking the butane and butylene releases lots of hydrogen that have a tendency to reattach themselves to the very reactive butadiene, undoing the whole point of the exercise. By adding oxygen to the reactor, most of the hydrogen reacts with it to form water.

$$C_4H_8 + \frac{1}{2}O_2 \rightarrow CH_2{=}CH{-}CH{=}CH_2 + H_2O$$

Reducing the excess hydrogen also permits longer residence time, (0.25 to 0.50 seconds) and lower operating temperatures.

Recovery by extractive distillation

When butadiene is produced in olefins plants or in refinery crackers, they come mixed with relatively large volumes of the other C_4 family. Sometimes the other C_4's need not be separated from each other, for example if they are going to be used for alkylation plant feed. In that case, the butadiene can be separated from the other C_4's by extractive distillation. This process uses a solvent that will preferentially dissolve butadiene, ignoring the other components in the stream.

The C_4 stream is fed to the middle of a fractionator, and a high boiling point solvent is fed at the top. The solvent, as it works its way down, strips out the butadiene as the C_4 vapor works its way up the column. The solvent and butadiene come out the bottom and can easily be split in a second column. Two popular high boiling point solvents are N-methylpyrrolidone (NMP) and Dimethylformamide (DMF). The chapter on benzene has more details on the extractive distillation process.

End-use markets

Butadiene's two double bonds make it very reactive. It readily forms polymers, reacting with itself to form polybutadiene. It's also used as a comonomer to make styrene-butadiene rubber (SBR), polychloroprene, and nitrile rubber. These are all forms of synthetic rubber and account for about 75% of the butadiene consumed. The largest share of them is on highway vehicles—truck and car tires, hoses, gaskets, and seals.

Some of the nonrubber applications are as a chemical intermediate to make adiponitrile and hexamethylenediamine, precursors to making Nylon 66 whose primary application is carpeting. Other nonrubber applications are styrene-butadiene latexes for paper coatings and carpet backing, and acrylonitrile-butadiene-styrene (ABS) resins for plastic pipe and automotive/appliance parts.

ISOBUTYLENE

The isobutylene in the C_4 stream generally ends up in one of four places: a refinery alkylation plant (covered further below), an MTBE plant, a polymerization process, or in gasoline. The first three are methods of removing isobutylene from the C_4 stream; the fourth is the default—the isobutylene just follows the other butanes and butylenes to the gasoline blending pool.

The major use of isobutylene as a separated petrochemical has become MTBE, a gasoline-blending component with two meritorious attributes. Like isooctane, it has a high octane number, plus it has oxygen in its molecular structure. While the former was enough to get interest in MTBE started, the latter had increasing appeal due to its environmental implications. The presence of the oxygen in the MTBE molecule, as shown in Figure 6–3, facilitates complete combustion of gasoline in a vehicle. That eliminates from the tail pipe exhaust almost all of the unburned hydrocarbons which are precursors to ozone and therefore a nasty air pollutant.

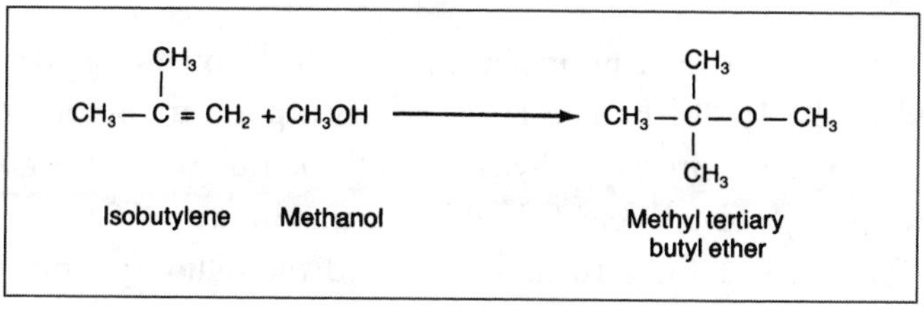

$$CH_3 - \overset{\overset{\displaystyle CH_3}{|}}{C} = CH_2 + CH_3OH \longrightarrow CH_3 - \overset{\overset{\displaystyle CH_3}{|}}{\underset{\underset{\displaystyle CH_3}{|}}{C}} - O - CH_3$$

Isobutylene Methanol Methyl tertiary butyl ether

Fig. 6–3 MTBE reaction

Chapter 13 covers the details of making MTBE. The process has a convenient attribute when it comes to C_4 hydrocarbons. The mixed stream can be fed to it *in toto*. The isobutylene is selectively reacted out.

Polymerization

Similarly, the polymerization process will pull the isobutylene selectively out of the C_4 stream. Polyisobutylene is used mainly as a viscosity index improver in lubricating oils and in caulking and sealing compounds. Some of the low molecular weight polyisobutylenes are particularly suited for use in the construction field because it doesn't solidify. They remain a tacky fluid and when properly formulated with clay fillers, etc., take on the properties of a sticky, putty-like substance.

The polymerization process is a low temperature catalytic reaction. The type of polymer produced is strongly affected by the reaction temperature. Low temperatures give low molecular weight polymers, the kind useful in caulking compounds and as a viscosity index improver for motor oils.

Solvent extraction. Isobutylene can also be segregated by extractive distillation in the same way as butadiene. In this case, the solvent is cold sulfuric acid. One problem occurs if there is any butadiene left in the stream—sulfuric acid will cause it to polymerize. But if the butadiene has been first extracted, a 99⁺% isobutylene stream can be recovered.

High purity isobutylene is used in numerous applications beside the polyisobutylene just mentioned: butyl rubber, oxo alcohols, tertiary butyl alcohols, di- and tri-isobutylene and methyl methacrylate.

BUTENE-1

The demand for high purity butene-1 (called polymer grade) rapidly developed in the 1970s. Butene-1 was always a popular comonomer with ethylene in high density polyethylene (HDPE). But the rapid growth of linear low density polyethylene (LLDPE) starting in the 1970s increased the demand for butene-1 from 10 million toward the billion pounds-per-year category in the ensuing years.

Other minor petrochemical uses of butene-1 continue to be the manufacture of SBA (secondary butyl alcohol), MA (maleic anhydride), and butylene oxide.

The boiling points of butadiene, isobutylene, and butene-1 make it impractical to recover a high purity butene-1 stream without first removing the other two by methods other than fractionation, as covered previously. After that, the butene-1 still needs to be separated from the other C_4's, but that can be done by fractionation. That's still an expensive proposition because the boiling temperatures of isobutane, normal butane, and butene-2 are not all that different. An alternate route, molecular sieve adsorption, works well.

Butene-1 separation processes

The applications of butene-1 usually require very low levels of isobutylene and butadiene. Sometimes an extra reactor in the MTBE plant is added to get the isobutylene content down from the typical 2.0% to a 0.2% level. Small amounts of butadiene are removed by hydrotreating the stream over a catalyst, which converts the butadiene to butene-2 and maybe some butane.

Distillation. The distillation method of separating butene-1 is difficult. It requires a column with more than 100 trays operating under a reflux ratio of about 150:1 (Chemical engineer jargon—it means a very tall column, with lots of recycle—very energy intensive.)

Adsorption. The second technology is selective adsorption. The use of molecular sieves was discussed in the chapter on xylenes. Molecular sieves, you will recall, are crystals with millions of pores, all of a uniform size or shape. In this process, a sieve with pores that will fit only butene-1 is used.

The process runs on a cycle. First the C_4 stream is fed to a vessel packed with the molecular sieve. The butene-1 molecules start to fill up the sieve's pores. After a while, when the pores are about saturated, the feed is cut off. Another liquid, the desorbent, is flushed back through the vessel, and the butene-1 is washed out of the sieves. The desorbent is selected so that after it picks up the butene-1 from the sieve, it can easily be separated from the butene-1 by fractionation. The key, of course, is to use a desorbent with a boiling temperature a good distance away from butene-1's. Any run-of-the-mill hydrocarbons that fit in this criterion are suitable.

BUTENE-2

In the chapter on olefins plants, in the section on propylene, a route to making propylene involved butene-2. In this process, called *metathesis*, ethylene and butene-1 are passed over a catalyst, and the atoms do a musical chair routine. When the music stops, the result is propylene. The conversion of ethylene to propylene is an attraction when the growth rate of ethylene demand is not keeping up with propylene. Then the olefins plants produce an unbalanced product slate, and producers wish they had an "on-purpose" propylene scheme instead of just a coproduct process. The ethylene/butene-2 metathesis process is attractive as long as the supply of butylenes holds out. Refineries are big consumers of these olefins in their alkylation plants, and so metathesis process has, in effect, to buy butylene stream away from the gasoline blending pool.

ALKYLATION

At the end of the line, after all the high value applications have finished with the C_4 hydrocarbons, is the alkylation plant. Most of the butene-2 ends up there, but butene-1 and isobutylene can as well. Butadiene will mess up the process, causing foam and hydraulic problems.

The term alkylation generally applies to the addition of an olefin to another organic compound. In the case of a refinery alkylation plant, the olefins are propylene or butylenes and the organic compound is isobutane. The result is C_7 or C_8 hydrocarbons called alkylate. Because they have the branch that comes along with the isobutane, they have high octane numbers and are good motor gasoline blending components. The alkylation of butene-1 with isobutane produces isooctane, which has an octane rating of 100. That number is not just a random occurrence. Isooctane is the compound that is used in its pure form to set the definition of octane rating.

The process

Alky plant feed comes in the form of the cracked gas streams, the C_3's and C_4's from a cat cracker or coker. The paraffins are included along with the olefins but they will not participate in the chemical reaction. The cracked gases are fed to a reactor, together with an excess of isobutane (about a 12:1 ratio). The reactors contain cold sulfuric or hydrofluoric acid that acts as a catalyst. Active mixing along with a residence time of 15–20 minutes results in reaction of the olefins with the isobutane. The propane and normal butane that is typically present with mixed C_3 or C_4 streams are unaffected by the process and just float on through. Distillation at the tail end of the plant easily separates the C_7 or C_8 alkylate from the propane, normal butane, and the unreacted isobutane, which is recycled to the reactor. Quality-wise, the alkylate made from butenes is better (higher octane) than that from propylene. (*See* Figure 6–4.)

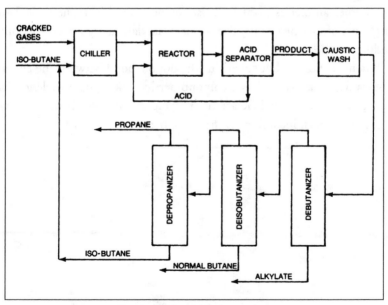

Fig. 6–4 Refinery alkylation plant

Chapter 6 in a nutshell ...

The C_4 family includes normal and isobutane, C_4H_{10}, which have only single bonds, normal and isobutylene; butenes, C_4H_8, which each have one double bond; and butadiene, C_4H_6, which has two double bonds. The reactivities and chemical versatility of these three groups are roughly related to the number of double bonds.

The source of these compounds is varied. The butanes are found naturally in crude oils and natural gas. They, plus the olefins, are products of various refinery processes and of olefins plants. They are separated by fractionation, except for butadiene and isobutylene, which are sometimes recovered by extractive distillation. They all vaporize at room temperature, so they are handled in closed, pressurized systems.

The butanes are used as gasoline-blending components. Normal butane is sometimes an olefins plant feed. Isobutane is used in refinery alkylation plants with propylene or butylene to make alkylate, a high-octane gasoline-blending component.

Butene-1, normal butylene with the double bond between the end and the second carbon, is used as a comonomer in making polyethylene. Polybutylene and polyisobutylene are the polymers. Butadiene is used to make complex polymers, including synthetic rubbers.

EXERCISES

1. Why do some people convert butylenes to butadiene but others convert butadiene to butylene?

2. Which do you suppose is the swing supply of butadiene:
    a. olefins plant
    b. refinery
    c. butane dehydrogenation
    d. butylene dehydrogenation
    e. Swinging Sal's butadiene shop

    Describe how you'd use the following processes to handle the C_4 stream in the heavy liquids cracker in Figure 5–3:
    a. extractive distillation
    b. fractionation
    c. adsorption
    d. polymerization
    e. dehydrogenation
    f. alkylation

COMMENTARY 1 ◇━━━━━━━━

"This is not the end.
It is not even the beginning of the end.
But it is, perhaps,
the end of the beginning."

**Winston Churchill, 1874–1965
Commenting on November 10, 1942
about the victory at Alamein**

REVIEW

After five chapters of product and process descriptions, you might be dismayed to find out it all can be summarized in one diagram (see following figure. At least the essential parts are there, and they can provide you quick reference.

The petrochemical products from olefins plants are ethylene, propylene, C_4's (butanes, butylenes, and butadiene) and a stream containing the BTXs. Refinery cat crackers produce propylene and C_4's. They produce some ethylene, but often it is not recovered.

The propylene from olefins plants is usually chemical grade; from cat crackers, refinery grade. Both can be upgraded to polymer grade by fractionation.

The diagram doesn't show the ethylene and propylene made by metathesis of methanol or the propylene made by catalytic reaction of ethylene and butylene or by dehydrogenation of propane. The volumes are small, and besides, it would make the diagram too messy.

Refinery cat reformers produce a reformate stream with aromatics. That stream, with or without the benzene-laden stream from the olefins plant, can be split apart in the various processing schemes in the BTX recovery facility.

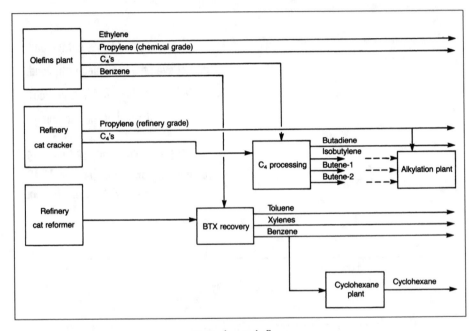

Basic chemicals flow

The C_4's can be recovered in numerous ways to make petrochemical feeds. Sometimes everything but the butadiene ends up in an alkylation plant that makes gasoline blend stock.

Benzene can be hydrogenated to cyclohexane directly.

FOREWORD

The next five chapters cover the "Mutt and Jeff" petrochemicals (or "the Gemini twins" if you're too young to remember who Mutt and Jeff were). They're grouped together because they come in pairs. You don't have one without the other. The metaphor is accurate for cumene and phenol, ethyl benzene and styrene, and ethylene dichloride and vinyl chloride. It's a little strained with ethylene oxide and ethylene glycol and with propylene oxide and propylene glycol because there are other things you can do with the oxides than make the glycols. But there's little other use for cumene, ethyl benzene, and ethylene dichloride than to use them to make their buddies.

Anyway, these 10 chemicals are the first- and second-generation derivatives of the basic building blocks and are important commodity petrochemicals.

CHAPTER 7 ⬡

Cumene and Phenol

"Every man serves a useful purpose:
a miser, for example,
makes a wonderful ancestor."

Lawrence J. Peter

The only reason petrochemical companies make cumene is to use it to make phenol. There are other ways to make phenol, but not much other commercial use for cumene.

CUMENE

During World War II, isopropyl benzene, more commonly and commercially known as cumene, was manufactured in large volumes for use in aviation gasoline. The combination of a benzene ring and an iso-paraffin structure made for a very high octane number at a relatively cheap cost. After the war, the primary interest in cumene was to manufacture cumene hydroperoxide. This compound was used in small amounts as a catalyst in an early process of polymerizing butadiene with styrene to make synthetic rubber. Only by accident did someone discover that mild treating of cumene hydroperoxide with phosphoric acid resulted in the formation of phenol and acetone. Serendipity is not uncommon in the discovery process involving petrochemicals.

In more recent vintages of the cumene manufacture processes, zeolyte catalysts permit going directly to cumene from the same two feeds, benzene and propylene. The introduction of catalytic distillation has even further improved the process economics, a thing that delights the manufacturers.

The phosphoric acid process

The reaction of benzene with propylene produces cumene (*see* Figure 7–1), but a catalyst must be present to make the reaction go. The chemistry is such that the benzene-propylene bond will be at the middle carbon of the propylene molecule, hence, the name isopropylbenzene. Note that there is also a transferal of hydrogen from benzene to the propylene.

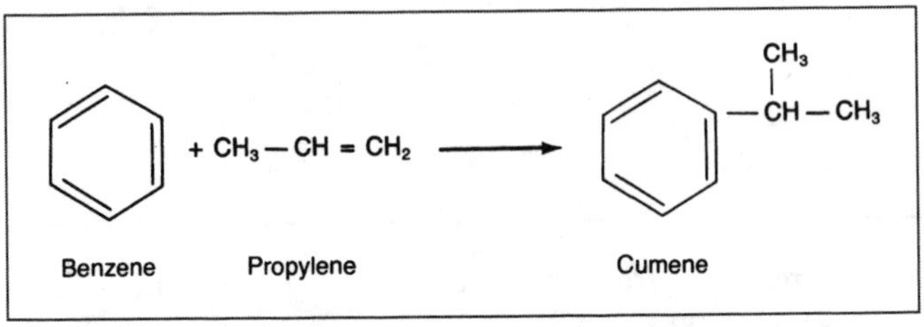

Fig. 7–1 Benzene-propylene route to cumene

The reaction can be carried out with the benzene and propylene in either the liquid or vapor phase; the more common process is vapor phase, carried out at about 425°F and 400 psi.

The process diagram in Figure 7–2 shows propylene and benzene being fed directly to the reactor. Chemical grade propylene with 6–10% propane is used because the presence of the propane does not affect the reaction. A front-end depropanizer can be used if a refinery grade propylene stream is used. That reduces the extra pumping costs and dilution effects of the propane.

The reactor is a vessel with beds of solid catalyst. Most commercial processes use a catalyst called kieselguhr, which is phosphoric acid deposited on a silica/alumina pellet. Because of the weight of the pellets, supported beds at multiple levels in the vessel are used so the bottom layers won't be crushed.

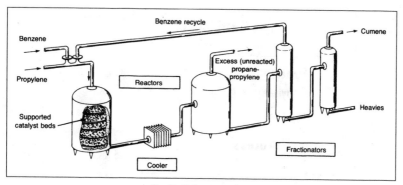

Fig. 7–2 Cumene plant

Two vessels are used for the reaction, and for two reasons. First, the reaction is exothermic and in a fixed catalyst bed, and one way to control the temperature is to take out the streams being processed and cool them down. The second reason is that the second reactor also is used as a fractionator, venting the unreacted propylene and the propane part of the chemical grade propylene from the benzene/cumene mix.

Excess benzene is always used in the reactors, also for two reasons. First, the benzene acts like a heat sponge, mitigating the rate at which the temperature increases due to the exothermic reaction. Second, excess benzene helps eliminate some of the undesirable side reactions that can take place, mainly the formation of di- or tri-isopropyl benzene (benzene hooking up with two or three propylenes) or other miscellaneous compounds.

The streams coming out of the reactors will be a mixture of the excess benzene and the product—cumene. A fractionating column is used to separate the two, permitting the benzene to be recycled. A final fractionator takes the cumene overhead; any of those miscellaneous compounds accidentally formed in the process go out the bottom.

Cumene made in this manner is about 99.9% pure. The cumene yield, *i.e.*, the percent of benzene that ends up as cumene, is about 95%. About 5% of the benzene ends up as part of the heavies. Conversion of propylene is a little lower, about 90%, particularly if there's no depropanizer up front to which the unreacted propane/propylene from the second reactor can be recycled.

Material Balance

Feed:

Benzene	681 lbs.
Propylene	387 lbs.

Product:

Cumene	1000 lbs.
Heavies	48 lbs.

The catalytic distillation process using zeolyte

In a Texas two-step that has led to a more economical route for cumene, new catalysts combined with a novel processing scheme has reduced both operating costs and increased the yield of cumene from its benzene and propylene feedstocks. In Figure 7–3, the main reaction takes place in a catalytic distillation column. This piece of apparatus combines a catalyst-filled reactor with a fractionator.

Chemical grade propylene is introduced to the middle section of the column as a vapor and will rise up through the catalyst bed at the top of the column. Benzene is introduced at the top of the column as a liquid and trickles down through the catalyst bed, mixing in countercurrent flow with the propylene vapors. Zeolyte-based catalysts cause the direct alkylation of benzene with propylene forming cumene. (What's alkylation? The addition of an olefin group to another organic compound.) One of the side reactions also produces an appreciable amount of polyisopropyl benzene (PIPB) that will have to be dealt with in the other reactor in Figure 7–3.

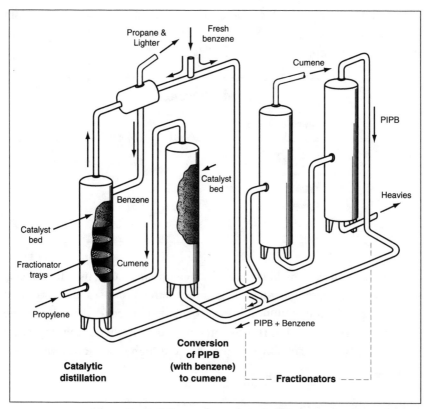

Fig. 7–3 Cumene by catalytic distillation

Both the cumene and the PIPB will continue down the column. The hot propylene and cumene vapors and the column trays will strip any remaining, unreacted benzene from the falling cumene/PIPB liquids, pushing it up the column, back to the catalyst bed where it will react with the propylene.

From the top of the column, the propane that always accompanies the propylene feed emerges, taking with it some of the benzene. In a flash tank, the propane is vented, and liquid benzene is recycled. From the bottom of the catalytic distillation column come the cumene, the PIPB, and some miscellaneous heavies that are separated in a fractionator to make cumene of 99.9% purity. The PIPB is separated in another column and fed to a second reactor with another zeolite catalyst bed. In there the PIPB reacts catalytically with benzene to produce additional cumene that is fed to the bottom of the catalytic distillation column to join the cleanup scheme. In that way, the yields get bumped up to 99.5–99.8%.

Material Balance	
Feed:	
Benzene	652 lbs.
Propylene	352 lbs.
Product:	
Cumene	1000 lbs.
Heavies	4 lbs.

Occasionally you might come across a compound called pseudo-cumene, which is a benzene ring connected to three methyl groups. This compound is an isomer of cumene known as 1,2,4-trimethyl benzene. Pseudo-cumene is a starting material for the manufacture of trimilletic anhydride, an important ingredient in alkyd resin paints and high temperature aerospace polyimide resins.

Commercial aspects

Cumene is a colorless liquid, soluble in benzene and toluene and insoluble in water. It can be shipped in tank cars, tank trucks, barges, and drums. The flash point is high enough that it is not considered a hazardous material, and no DOT red shipping label is required.

Cumene Properties	
Molecular weight	120.19
Freezing Point	-140.8°F (-96°C)
Boiling Point	306.5°F (152.5°C)
Specific gravity	0.8632 (lighter than water)
Weight per gallon	7.19 lbs/gallon

The grades used in commerce are Technical (99.5% concentration), Research (99.9%), and Pure (only trace impurities). About 97% of production is presently Technical grade.

PHENOL

Phenol has been used for decades in the medical field as an antiseptic under its alias, carbolic acid, and at one time as a preservative of human organs under the name creosote (from the Greek *kreos*, flesh; and *sogein*, to preserve). The name *creosote* eventually became associated with the wood preservative, but phenol remains the principal ingredient in this product.

The early sources of phenol were the destructive distillation of coal and the manufacture of methyl alcohol from wood. In both cases, phenol was a by-product. Recovered volumes were limited by whatever was made accidentally in the process. Initial commercial routes to "on-purpose" phenol involved the reaction of benzene with sulfuric acid (1920), chlorine (1928), or hydrochloric acid (1939); all these were followed by a subsequent hydrolysis step (reaction with water to get the -OH group) to get phenol. These processes required high temperatures and pressures to make the reactions go. They're multistep processes requiring special metallurgy to handle the corrosive mixtures involved. None of these processes is in commercial use today.

In 1952, a technological breakthrough was found: the cumene oxidation route. It was much cheaper, and it quickly proliferated. It is now the primary route, accounting for almost all of U.S. production.

The cumene oxidation process

This two-step process involves oxidation of cumene to cumene hydroperoxide, which decomposes with the help of a little dilute acid into phenol and acetone. In the first step, cumene is fed to an oxidation vessel (as shown in Figure 7–5), where it is mixed with a dilute aqueous sodium carbonate solution (soda ash with a lot of water). A small amount of sodium stearate is added, and the whole mixture becomes an emulsion.[1]

The purpose of the cumene emulsion is to permit good contact of the cumene with oxygen. The oxygen is introduced as air in the bottom of the vessel and bubbled through the emulsion. As it does, the cumene converts to cumene hydroperoxide, as shown in step 1 in Figure 7–4.

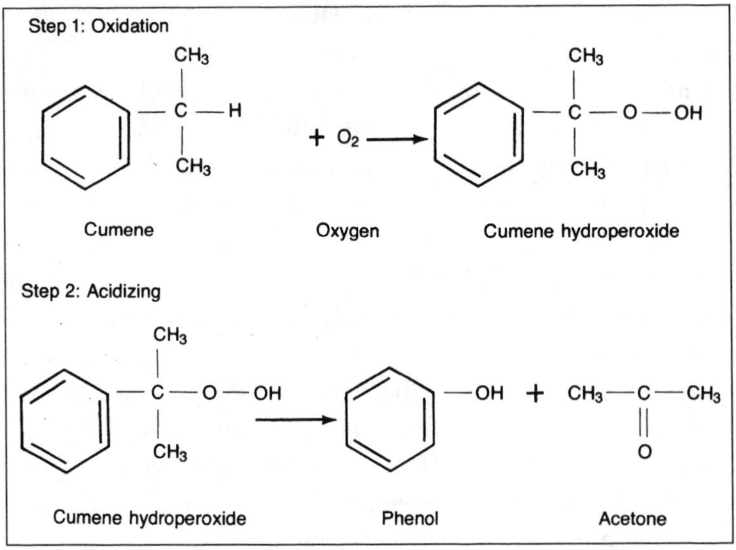

Fig. 7–4 Cumene-to-phenol process

This chemical reaction, like most oxidations, is exothermic, so it generates heat and is susceptible to a runaway reaction, sometimes called the "begets." Rapidly increasing temperatures begets an increased rate of reaction, which begets more rapidly increasing temperatures, which begets . . . The presence of the excess water sponges up some of the heat and reduces the risk of the begets.

To further control the runaway risk, the reaction temperature is kept at 230–250°F by regulating the reactor flow-through rate. At that temperature, only about 25% of the cumene is converted to phenol. The stream coming out of the bottom of the oxidizing vessel is 25% cumene hydroperoxide and 75% unconverted cumene.

At the top of the vessel (*see* Figure 7–5) is the necessary plumbing to take off the nitrogen content of the air, which just passes through untouched, plus any excess oxygen not used in the reaction.

The bottom stream is fed to a fractionator to split out the unreacted cumene in order to recycle it to the oxidizing vessel. The cumene hydroperoxide, now concentrated to about 80%, is fed to another vessel for the second reaction step. The chemical trick here is to chop out everything in between the benzene ring and the -OH group, as shown in step 2 of Figure 7–4. The use of dilute sulfuric acid does the job by initiating an unusual decomposition involving actual migration of the benzene ring around the cumene hydroperoxide molecule. The decomposition and the resulting products were a complete surprise when they were discovered. Again, stepping in a bucket of serendipity played a role in generating progress in this chemical.

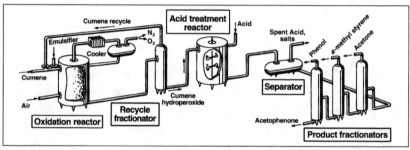

Fig. 7–5 Phenol plant

To facilitate the reaction, the mixture is stirred vigorously. It takes place at 150–175°F and 25–50 psi, so the conditions here, like in the oxidizer, are not too severe or expensive.

The effluent from the acid treatment reactor is about 60% phenol, 35% acetone, plus some miscellaneous nits and lice, most of which are alpha-methyl styrene and acetophenone. The effluent is passed through a separator where the acid, water, and salts drop out. The balance of the processing is a series of distillation columns that split out the various streams.

The alpha-methyl styrene can be recovered as a product or catalytically treated with hydrogen and converted back to cumene for recycling. The acetophenone has some commercial use in pharmaceuticals and at one time was used to make ethylbenzene. A high purity phenol is sometimes made by a crystallization step, since phenol freezes at about 109°F. With alpha-methyl styrene recycled, the ultimate yield is about 97%.

Material Balance	
Feed:	
Benzene	1310 lbs.
Propylene	352 lbs.
Sodium carbonate	Small
Sulfuric acid	Small
Product:	
Phenol	1000 lbs.
Acetone	615 lbs.
By-products	43 lbs.
Unused oxygen	45 lbs.

Other routes. Alternate process technologies for making phenol avoid the cumene route. A few plants have used toluene as a feed, oxidizing it over a cobalt catalyst to give benzoic acid. That is followed by a reduction (removal of oxygen atom) to give phenol and carbon dioxide.

Two other, more recently popular routes are shown in Figure 7–6. In the first, benzene is hydrogenated to cyclohexane, followed by a partial oxidation to cyclohexanol. The cyclohexanol is then dehydrogenated to phenol.

In the second route shown in Figure 7–6, benzene is converted directly to phenol by a catalytic reaction with nitrous oxide. Neither of these routes contribute an appreciable volume to phenol supply because they are new. However, an attractive feature of both is the fact that they produce no by-product acetone, a recurring Achilles heel for the phenol suppliers. For example, when acetone is short, running the phenol-acetone plants hard to make more acetone results in a glut of phenol and poor prices. When acetone is long and commands low prices itself, the phenol-acetone plant margins are depressed, making it difficult to warrant running it at high rates to supply enough phenol. While any plant with coproducts has this problem, manufacturers of phenol seem to complain about its recurrence more than any others.

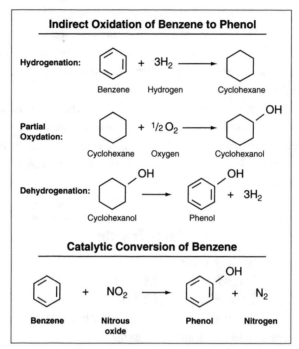

Fig. 7–6 Alternate routes to phenol

Commercial aspects

Uses. The major applications of phenol are phenolic resins, Bisphenol A, and caprolactam. The reaction of phenol with formaldehyde gives liquid phenolic resins (used extensively as the adhesive in plywood) and solid resins (used as engineering plastics in electrical applications). In powder form, the phenolic resin can be molded easily and are completely nonconductive. These phenolic resins or plastics can be found in panel boards, switchgears, and telephone assemblies. The agitator in your washing machine is probably a phenolic resin.

Phenol is also used to manufacture several important monomers. Bisphenol A, a phenol derivative, is used to make very strong polycarbonate plastics and epoxy resins (the kind you buy in two tubes and mix to make glue). Other applications of epoxy resins include paints, fiberglass binder, and construction adhesives.

About half the caprolactam is made from phenol. (The other half comes from cyclohexane.) Caprolactam is an intermediate step in making Nylon 6.

Other miscellaneous derivatives of phenol include non-ionic detergents, aspirin, disinfectants (pentachloro phenol), adipic acid (a Nylon 66 intermediate), and plasticizers.

Properties. Phenol is a solid at room temperature and is usually handled as a powder. In its pure form, it is white in color but exposure to sunlight or air will cause it to turn reddish. Phenol is and acts like acid. It burns, it's corrosive, and it has an odor and taste that will knock you over—literally. It's a Class B poison.

Phenol Properties	
Molecular weight	120.19
Freezing point	109.4°F (43°C)
Boiling point	359.2°F (181.8°C)
Specific gravity	1.071 (heavier than water)
Weight per gallon	9.0 lbs/gallon

Phenol can be shipped in liquid form in lined tank cars or tank trucks or in galvanized drums. It's imperative that it be handled in closed systems, as it will absorb water from the atmosphere. In powder form, phenol will absorb enough water from the atmosphere to turn itself into a liquid.

The powder form of phenol is usually traded either as a U.S.P. (98% minimum) or a C.P. or synthetic grade (95% minimum).[2] In the liquid form, the commercial grades are 90–92% purity and 82–84% purity.

Chapter 7 in a nutshell ...

Cumene, $C_6H_5CH(CH_3)_2$, is a benzene ring with a unique functional group hung on it in the place of a hydrogen atom. It is made by reacting benzene with propylene. Fifty years ago, it was used primarily as a high-octane aviation gasoline component, but it is now used almost entirely as feed to coproduct manufacture of phenol and acetone.

Phenol, C_6H_5OH, is a benzene ring with a hydroxyl group, -OH, in place of a hydrogen. That makes it a member of the alcohol family. Most phenol is made by the oxidation route. At room temperature phenol is a solid but is corrosive like an acid. It is used to make phenolic resins and to make Bisphenol A (feed for epoxy and polycarbonate resins) and caprolactam (feed for Nylon 6).

EXERCISES

1. The Cutthroat Phenol Company is thinking about starting up and selling out their 50 million pound per year phenol plant. The market for the various feedstocks and products is as follows. (F.O.B. the Cutthroat plant):

Benzene	$ 1.60/gal
Propylene	0.25/lb
Oxygen	0.50/lb
Acetone	0.60/lb
By-products	0.10/lb

   Their cumene plant operating costs are $0.20/lb of feed; the phenol plant operating costs are $0.25/lb of phenol.

   How much should Cutthroat charge for phenol if they want to make $0.02/lb.

2. What's wrong with Cutthroat's logic?

Endnotes for Chapter 7

[1] Emulsions abound in everyday use. Mixing flour and water with meat juices make the emulsion called gravy. Putting soap powder in a washing machine causes the dirt in clothes not only to be removed but become suspended in the water—an emulsion. The dirt particles remain in this state until they're rinsed away by draining the washing machine. Seventy-five years ago the emulsion was lye. Before that, washing was mechanical—rubbing on a washboard, or even a rock.

[2] U.S.P. (United States Pure) and C.P. (Chemical Pure) are nomenclature from the pharmaceutical industry, the former indicating a grade suitable for human consumption or for manufacture of a consumable.

CHAPTER 8 ⬡

Ethylbenzene and Styrene

Diogenes the wise crept into his vat
and spoke:
"yes, yes, this comes from that."

Wilheim Busch Diogenes, 1832–1908
(inventor of the cartoon strip)

ETHYLBENZENE

Ethylbenzene is to styrene what cumene is to phenol. The only reason you want to make ethylbenzene is so you can make styrene. Its destiny is tied to styrene consumption. Most ethylbenzene (EB) is made by alkylating benzene with ethylene, as shown in Figure 8–1.

A small amount of EB is present in crude oil and also is formed in cat reforming. You might recall from Chapter 3 that there is only a 4°F difference between the boiling points of EB and para-xylene. Consequently, a superdistillation column is needed for the separation. In process engineers' terms, it would have about 300 theoretical trays, be about 200 feet tall, and even then have a high reflux ratio to accomplish the separation. All this is necessary because the EB stream must be quite pure to be used for styrene manufacture.

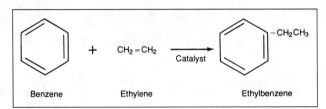

Fig. 8–1 Alkylation of benzene with ethylene to form ethylbenzene

The technology

Alkylation of benzene is old technology. The French chemist, Charles Friedel, with his American partner, James Crafts, in 1877, stumbled (almost literally) across the technique for alkylating benzene with amyl chloride ($C_5H_{11}Cl$). The use of a metallic catalyst, in this case aluminum, was the key. The Friedel-Crafts reaction is classical and remains a principal route for alkylating benzene with ethylene to make EB.

The Friedel-Crafts reaction has one major drawback. It doesn't stop at the mono-substitution stage. That is, the catalyst works so well, that the benzene will pick up two, three, or more ethylene molecules, forming diethylbenzene, triethylbenzene, or higher polyethylbenzenes. (*See* Figure 8–2.) The problem is that chemically it's easier to alkylate EB than it is benzene. One way to control the problem is to carry out the reaction in the presence of a large excess of benzene. When an ethylene molecule is in the neighborhood of one EB molecule and 20 benzene molecules, chances are that the ethylene will hook up with benzene, even though it prefers EB.

The other controllable variable is the operating conditions. Certain temperature and pressure levels will favor the benzene and not EB alkylation—not exclusively, but predominantly. In fact, these variables can be set to favor the di- and triethylbenzenes to give up an ethyl group to benzene to give EB. That process is called transalkylation and is shown in Figure 8–2.

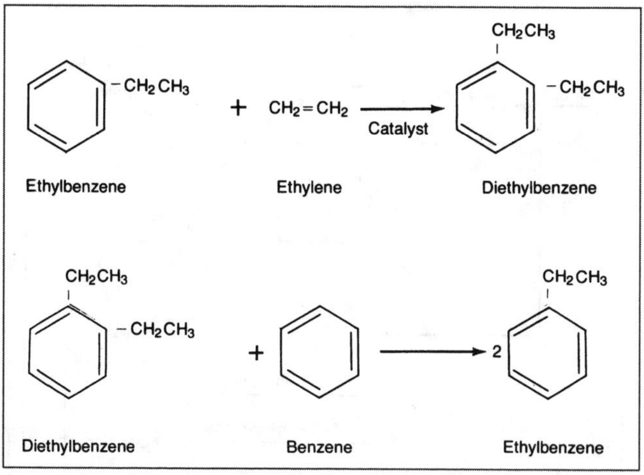

Fig. 8–2 Formation of diethylbenzene and transalkylation to ethylbenzene

Processes

The alkylation of benzene with ethylene is done commercially either in the liquid phase or the vapor phase. The older liquid phase operation is shown in Figure 8–3. The catalyst is anhydrous aluminum chloride (anhydrous meaning completely water-free). At temperatures of 300–400°F and pressures of 60–100 psi, the reaction time is about 30 minutes, so the reactor must be large enough to accommodate this long residence time. At higher temperatures, alkylation occurs in a single homogeneous phase (homogeneous meaning only one, gaseous or liquid), and less catalyst is required. Since spent aluminum chloride catalyst, like a number of other spent catalysts, is an annoying disposal problem for the producers, the higher energy expenses have turned out to be a reasonable price to pay to reduce the nuisance.

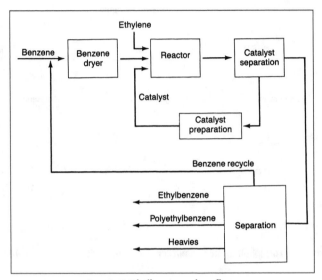

Fig. 8–3 Ethylbenzene plant flows

Sometimes a catalyst promoter or accelerator, ethyl chloride, is added to the feed to speed up the reaction. The ethyl chloride actually works on the aluminum chloride catalyst, not the reactants. It's like offering a supervisor a bonus. He doesn't do any more work, but he gets more work done.

The effluent stream leaving the reactor is cooled and then treated with caustic (sodium hydroxide) and water to remove the catalyst. The cleaned up stream then contains about 35% unreacted benzene, 50% EB, 15% poly-ethylbenzene (PEB), and a small amount of miscellaneous heavy materials.

The separation section uses three columns to separate the effluent stream into four components: unreacted benzene that is recycled, EB, PEB, and heavier by-products.

The PEB can be and usually is fed to a separate reactor (not shown) where it reacts with more benzene at 250–300°F in the presence of an aluminum chloride catalyst to produce additional EB via the transalkylation route. The catalyst is removed from the reaction mixture before it is passed into the separation section.

Usually, the benzene is water-washed before it is returned to the feed line; hence, the need for the dryer, which is often an azeotropic distillation column. See Chapter 3, the section on toluene, to read about azeotropic distillation. The presence of any water in the reactor would cause lots of undesirable side reactions to occur.

With full recycle of the benzene and processing of the PEB, the EB yield (yield being the percent of feed that ends up as the targeted product) is about 99%, based on the ethylene and benzene feed. (For a discussion of the difference between yield and conversion, see Appendix A.)

Material Balance

Feed:

Benzene	743 lbs.
Ethylene	267 lbs.
Catalyst	small amount
Ethyl chloride	small amount

Product:

Ethylbenzene	1000 lbs.
Polyethylbenzenes	10 lbs.

The alternate process, the vapor phase method, is carried out at higher pressures (450 psi) and temperatures (750–800°F), and hence, the vapor phase. Producers have been using a boron trifluoride catalyst but any trace water corrodes it unmercifully. Most have now switched to a crystalline aluminosilicate zeolyte catalyst, a more expensive but hardier catalyst. The newer catalyst is also noncorrosive and nonhazardous, cheaper to handle, and produces no waste streams to dispose of.

The flow diagram for the vapor phase looks about the same as Figure 8–3. But unlike the liquid phase process, in the reactor both alkylation and transalkylation take place simultaneously so there is no need for a separate reactor to convert PEB to EB. Virtually no PEB shows up as by-product.

The vapor phase process usually features dual reactors because the catalyst needs to be regenerated about every eight weeks. One reactor is off-line being regenerated while the other is operating.

Alternate EB technology

Another route to ethylbenzene is available for those remote places where olefin plants or refinery crackers are not nearby but a supply of ethane is—catalytic dehydrogenation of ethane to ethylene followed by its reaction with benzene to produce EB. The first of two steps in Figure 8–4 use a gallium zinc zeolyte catalyst that promotes ethane dehydrogenation to ethylene at 86% selectivity and up to 50% conversion per pass.

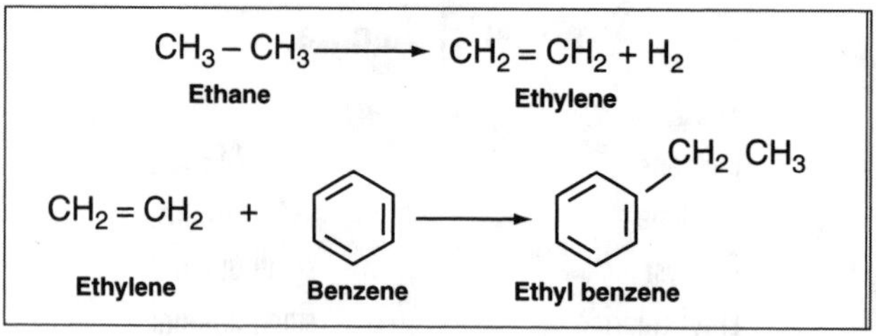

$$CH_3-CH_3 \longrightarrow CH_2=CH_2 + H_2$$

Ethane   Ethylene

$$CH_2=CH_2 \; + \; \text{Benzene} \longrightarrow \text{Ethyl benzene} \; (CH_2\,CH_3)$$

Ethylene   Benzene   Ethyl benzene

Fig. 8–4 Ethane/benzene route to ethylbenzene

In the second step, the crude ethylene plus benzene are passed over a silica mordenite catalyst at 650–700° F and atmospheric pressure to produce EB at 91% selectivity and 23% ethylene conversion.

The low conversion rates for both the ethane dehydrogenation and the ethylene-to-EB steps result in high capital costs for a world-scale plant. That limits the potential application of this process to boutique sites.

Handling

Since most of the EB is used for the manufacture of styrene, EB plants are usually found in close proximity to styrene plants. Very little EB is traded commercially or transported. A small amount of EB is used as a commercial solvent, mainly as a substitute for xylenes.

EB is not toxic like the xylenes, and it does not require the DOT red shipping label. EB has the same colorless appearance of the other BTXs as well as that characteristic sickly odor. Like the BTXs, it is insoluble in water.

STYRENE

The rapid growth of styrene after World War II was due to the widespread use of its derivatives, principally synthetic rubber and plastics. Styrene ought to be called one of the basic building blocks of the petrochemicals industry. But you get all mixed up with semantics because it's made up of two other basic building blocks, ethylene and benzene. Nonetheless, it is the most important monomer in its class.

You may wonder what makes styrene, an unlikely looking building block, so valuable. There is an explanation from the field of atomic chemistry about the special synergy that results from the combination of the benzene ring and ethylene. Both chemicals are each very reactive. But, while the relationship might intrigue you, the explanation is more complex than can be handled "in nontechnical language." In the end, the value of styrene comes from the ease of handling, the safe processing characteristics, and the low costs of the products that can be made from it.

The dominant share of styrene production comes from dehydrogenation of EB in plants like that shown in Figure 8–5. Some comes as a coproduct in propylene oxide/styrene plants. An even smaller amount is recovered from the gasoline fraction of olefins plants cracking heavy liquids.

Fig. 8–5 Polysar Ltd.'s styrene monomer plant at Sarnia, Ontario

Ethylbenzene dehydrogenation

The process of dehydrogenation of EB is shown in Figure 8–6. The process is similar to operations in an olefins plant in that dehydrogenation is done by mixing the feed with steam and cracking it in pyrolysis furnaces. However, the cracking products are more limited, primarily because of the use of a catalyst, iron oxide.

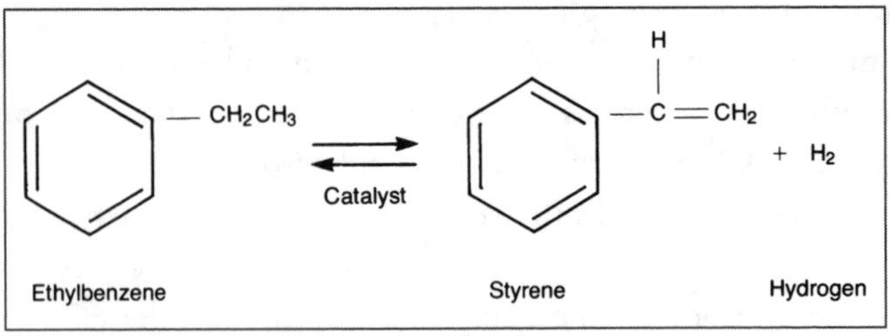

Fig. 8–6 Dehydrogenation of ethylbenzene to styrene

The dehydrogenation step, like all cracking processes, is endothermic (absorbs heat). Superheated steam, mixed with the EB, provides the heat and also performs two other important functions.

1. It reduces the pressure at which the reaction will take place.

   A mixture of EB and steam at, say, 1300°F can be contained at a lower pressure than EB by itself at that temperature. So what, you say?

   Well, the chemical reaction of EB cracking to styrene and hydrogen is reversible. Styrene easily hydrogenates back to EB. The reaction that is favored (both will always occur to some extent) will be determined by the operating conditions. In this case, higher pressures favor formation of EB because EB takes up less volume than the corresponding amount of styrene and hydrogen.

   Conversely, lower pressures favor formation of styrene. So the logic is that steam mixed with the EB permits cracking the hydrogen off at lower pressure and favors the styrene "staying cracked." (You may have noticed the chemical equation in Figure 8–6 has arrows going both directions. That's the chemist's notation for this reversibility.)

2. The steam also reacts with coke deposits on the iron oxide catalyst, forming CO_2, giving the catalyst a longer, more active lifetime. The onstream factor of the styrene plant is extended by reducing the shutdown frequency for catalyst regeneration or replacement.

In this process, conversion rate, the percent of feed that "disappears" in one pass through the reactor, is about 60%; the yield, the percent of feed that ends up as product, is about 90%. (Again, the difference between conversion and yield, if you want to know, is discussed in the Appendix.) The 10% yield loss results from cracking one of the carbon-carbon bonds along EB's ethyl group. Consequently, benzene and toluene make up most of the by-products. Another is polystyrene. The styrene, of course, is very reactive. That's why it's a building block. Though the reactor operating conditions minimize this reaction, some of the styrene molecules do join up to form styrene polymers.

In this process, it is very uncommon to detect any benzene rings breaking up. As in the olefins plant, the thermal stability of the benzene ring is demonstrated by its survival under these severe operating conditions, especially the high temperatures.

Material Balance	
Feed:	
Ethylbenzene	1132 lbs.
Catalyst	Small
Product:	
Styrene	1000 lbs.
Hydrogen	19 lbs.
By-products	113 lbs.

The process facilities

The EB entering the styrene plant is generally heated to the threshold cracking temperature (about 1100°F) in a heat exchanger[1]. The counter flow in the exchanger is the effluent from the second stage reactor, as shown in Figure 8–7. Because high temperatures are necessary in a styrene plant, energy conservation plays a big role in the plant design. While this is the only heat exchanger shown, recovery of waste heat is an intimate part of the process flow throughout the plant.

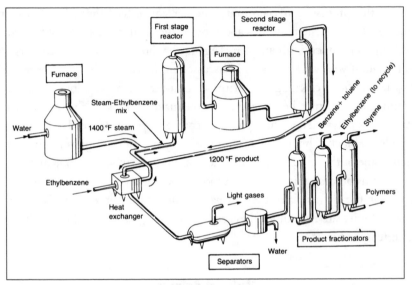

Fig. 8–7 Styrene plant

After heating, the EB is mixed with superheated steam and fed to the first stage reactor. Both the first and second stage reactors are packed with a catalyst of metal oxide deposited on an activated charcoal or alumina pellets. Iron oxide, sometimes combined with chromium oxide or potassium carbonate, is commonly used.

The actual reaction takes place at about 1150°F, but there is a temperature drop in the reactor as the dehydrogenation takes place. Reheating the stream in a furnace or exchanger is necessary before the stream is fed to the second stage reactor for a repeat performance.

The hot effluent is cooled in another heat exchanger, passing the heat off to the incoming fresh EB feed and maybe even further (not shown) with water to make steam. The cooled stream is then sent to separators where the light cracked gases that are unavoidably formed (H_2, CO, CO_2, CH_4, etc.)

and water are removed. The final product separations are done in a series of fractionators. The EB is recycled to the feed line; the polymers, very small in volume, are generally disposed of in residual fuel.

Even at ambient temperatures, styrene is likely to react with itself—very slowly, but steadily. For this reason, a small amount of polymerization inhibitor, about 10 ppm (parts per million) of para-tertiarybutyl catechol, a chemical whose name nobody wants to remember except its salesman, is added to styrene kept in storage. Since polymerization is promoted by higher temperatures styrene is usually stored in insulated tanks.

Alternate routes to styrene

A few plants are designed to produce styrene from EB but as a coproduct with propylene oxide (PO). In this process, EB is oxidized to a hydroperoxide (A in Figure 8–8) by bubbling air through the liquid EB in the presence of a catalyst. Hydroperoxides are, by their nature, very unstable compounds (one of the reasons that bleach, another hydroperoxide, works so well). So exposure to high temperatures has to be limited. The reactions are usually run at about 320°F and 500 psi pressure. Heat exchangers and multiple vessels are used to control the temperatures. Pressures are not critical in this process.

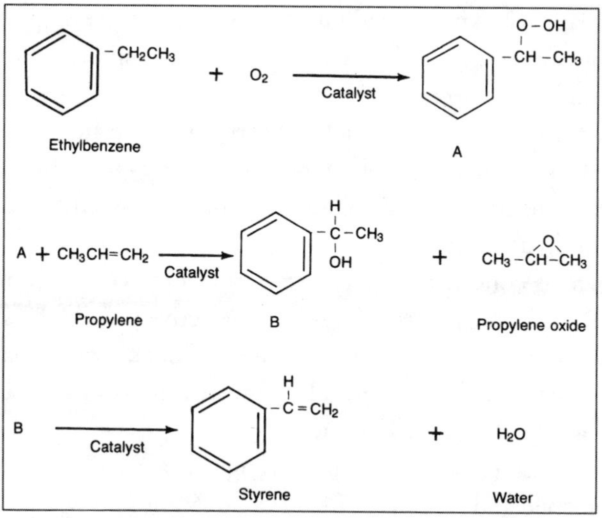

Fig. 8–8 Styrene and propylene oxide by ethylbenzene oxidation

The reactor effluent is distilled and unreacted EB is recycled. The EB hydroperoxide is then reacted with propylene at 250°F and pressure in the range of 250–700 psi in the presence of a metal catalyst to produce propylene oxide and methylbenzyl alcohol (*B* in Figure 8–7). The reactor mixture is separated by multiple fractionators. Unreacted propylene and EB are recycled. PO is recovered overhead. The methyl benzyl alcohol is easily dehydrated in the vapor stage at 450–500° F and 500 psi pressure over a titanium dioxide or silica gel catalyst to form styrene. Acephenone is one of the by-products.

The overall yield of styrene (the amount of EB that ends up as styrene) via this peroxidation process route is 90%.

The PO/styrene process is one of those coproduct operations where the economics get muddled easily. When styrene is in short supply and high-priced, PO from these plants look good. When PO is in long supply and low-priced, styrene from these plants looks anemic. The product managers of both products are rarely in the same mood.

Styrene from olefins plants

In the chaos that happens in a 10th of a second in the cracking furnaces of an olefins plant, sometimes a styrene molecule emerges. It happens often enough in plants cracking heavy liquids to warrant processing the gasoline fraction, pygas, to recover the styrene. The process is similar to the recovery schemes discussed in the aromatic chapters.

Pygas is fractionated to make a C_8 "heart cut," a stream with the styrene concentrated in it. This is usually done before the pygas is hydrotreated in preparation for using it as a gasoline blending component. Otherwise the styrene would hydrogenate as well.

The styrene concentrate is fed to a solvent recovery process or an extractive distillation process. The solvent selectively pulls the styrene out of the hydrocarbon mixture. The styrene raffinate, *sans* styrene, is sent back to be mixed with the pygas (although it can also be fractionated to pull out a high quality mixed xylene.)

The styrene-laden solvent is run through in a fractionator to produce solvent for recycle and styrene. A final fractionation step will get the purity of the styrene up to 99.9%.

A typical worldscale olefins plant producing a billion pounds a year of ethylene from heavy liquids can also yield up to 50 million pounds of styrene. Since the styrene is a coproduct, and the extraction costs are modest, the economics are very attractive compared to on-purpose styrene.

Commercial aspects

Uses. Plastics and synthetic rubber are the major uses for styrene. They account for the exponential growth from a few million pounds per year in 1938 to more than 8 billion pounds today. The numerous plastics include polystyrene, styrenated polyesters, acrylonitrile-butadiene-styrene (ABS), styrene-acrylonitrile (SAN), and styrene-butadiene (SB). Styrene-butadiene rubber (SBR) was a landmark chemical achievement when it was commercialized during World War II. The styrene derivatives are found everywhere—in food-grade film, toys, construction pipe, foam, boats, latex paints, tires, luggage, and furniture.

Styrene Properties	
Molecular weight	104.15
Freezing Point	-23.1°F (-30.6°C)
Boiling Point	293.4°F (145.2°C)
Specific gravity	0.9045 (lighter than water)
Weight per gallon	7.55 lbs/gallon

Handling. Styrene is a colorless liquid but tends toward a yellowish cast as it ages. It feels oily to the touch and smells like the aromatics compounds. Left alone at room temperature, styrene will eventually polymerize with itself to a clear glassy solid.

Technical grade styrene is 99% minimum purity. It is shipped, with a polymerization inhibitor in it, in standard tank cars or trucks. However, it has none of the severe handling precautions that benzene does.

Chapter 8 in a nutshell ...

Ethylbenzene, $C_6H_5C_2H_5$, belongs in the BTX family because it is a benzene ring with an ethyl group, $-CH_2CH_3$, attached in place of a hydrogen. It is made by reacting benzene and ethylene. Virtually all ethylbenzene is used to make styrene.

Styrene is a benzene ring that has a double bonded group attached that gives it the reactivity that makes it so useful. To convert ethylbenzene to styrene, quick, high temperature exposure in a cracking furnace is used. The ethylbenzene, fortunately, preferentially loses two hydrogens from the ethyl group, leaving a double bonded carbon.

EXERCISES

1. _____ is to phenol as EB is to _____.

2. By-products are important. What are two discussed in the last two chapters that could have important effects on product economics?

3. How many pounds of styrene could you make starting with 1000 pounds of benzene? How much ethylene would you need?

Endnotes for Chapter 8

[1] In a heat exchanger, the two streams trading heat never come in physical contact with each other. One stream is inside a set of pipes, or more commonly, exchanger tubes, that pass through a vessel containing the other stream. The heat exchange comes about by heat (only) passing through the tube walls. After all, if the streams were mixed to transfer the heat, they'd have to be reheated to separate them by fractionation.

CHAPTER 9 ⬡

Ethylene Dichloride and Vinyl Chloride

"Inventing is a combination of brains and material.
The more brains you use,
The less material you need."

Charles F. Kettering, 1876–1963
President, General Motors

This is the third chapter in a series of three where the products are like pancakes and batter. You can't make pancakes (or vinyl chloride) without making batter (or ethylene dichloride); there's not much else you can do with pancake batter (or ethylene dichloride); and you only use pancakes or vinyl chloride to make something else—breakfast or plastics. Finally, if you'll permit this painfully overbearing analogy to be extended once more, making vinyl chloride (or pancakes) from scratch is a lot easier now than it was 40 years ago.

The original manufacturing route to vinyl chloride (VC) didn't involve ethylene dichloride (EDC) but was the reaction of acetylene with hydrochloric acid. This process was commercialized in the 1940s, but like most acetylene-based chemistry in the United States, it gave way to ethylene in the 1950s and 1960s. The highly reactive acetylene molecule was more sensitive, hazardous, and eventually more costly than ethylene. The chemical engineers were happy to replace acetylene technology with the ethylene route. All the contemporary vinyl chloride plants now use ethylene and chlorine as raw materials.

Vinyl chloride is often called vinyl chloride monomer (VCM). The tag-on, monomer, from the Greek *mono*, meaning one, and *meros*, meaning part, is a convention used to contrast a chemical from its counterpart, the polymer polyvinyl chloride, PVC. Vinyl is the prefix for any compound that has the vinyl group, $CH_2=CH-$, in it. The root of the word is the Latin *vinum*, meaning wine, perhaps having something to do with a preoccupation of the discoverer.

THE PROCESS

VC is made by cracking EDC in a pyrolysis furnace much like that in an ethylene plant. That's one of the three reactions, shown in Figure 9–1, involved in the process. The other two have formidable names—chlorination and oxychlorination—but simple enough reactions—the addition of chlorine and the addition of oxygen and chlorine. What is a little complicated is the fact that the hydrogen chloride used to make the EDC in the first reaction comes from cracking EDC in the second. Sounds like a closed circle until you peel it back and examine it.

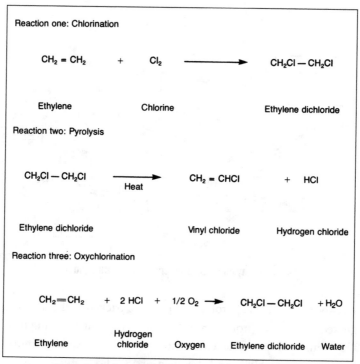

Fig. 9–1 Ethylene dichloride and vinyl chloride reactions

Figure 9–2 shows the plant with its three reactors. The pyrolysis furnace is in the middle. At the top of the figure, the basic feeds to the plant are shown—ethylene, chlorine, and oxygen. Ethylene and chlorine alone are sufficient to make EDC via the route on the left. The operation, call it Reaction One like Figure 9–1 does, takes place in the vapor phase in a reactor with a fixed catalyst bed of ferric (iron) chloride at only 100–125°F. A cleanup column fractionates out the small amount of by-products that get formed, leaving an EDC stream of 96–98% purity.

For Reaction Two, the purified EDC is passed through a dryer to remove water, then fed to a pyrolysis unit. The difference between EDC pyrolysis furnaces and those used for ethylene is the use of a catalyst.

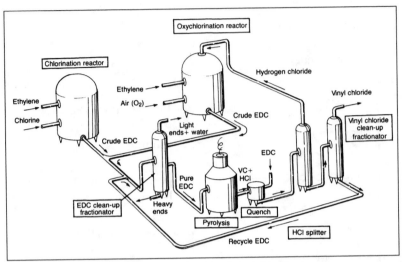

Fig. 9–2 Ethylene dichloride and vinyl chloride plant

The tubes in the EDC pyrolosis furnace are packed with charcoal pellets impregnated with ferric (iron) oxide. The EDC is pumped through at about 900–950°F and 50 psi. The conversion of EDC, i.e., how much of it "disappears," is about 50%, and the yield of VC, how much of the "disappearing" EDC gets converted to VC, is about 95–96%. (*See* the Appendix if you haven't yet read about the difference between yield and conversion.)

So not much else is formed. That's a contrast to ethylene manufacture, especially cracking the heavy liquids, where the by-products are abundant.

The hot effluent gas from the furnaces is quenched right away for the same reasons as ethylene furnace effluent is quenched—to stop the cracking at the optimum point. In this case, though, the quench liquid is cool EDC, not water.

When EDC cracks, one hydrogen and one chlorine on adjacent carbon atoms are sprung and find each other, forming hydrogen chloride gas. (Why one of each cracks off, and not two hydrogens or two chlorines is another mystery of atomic physics.) The cooled effluent is fractionated into three streams: hydrogen chloride, EDC, and the VC stream, which is sent to storage. The EDC, which is the unconverted pyrolysis feed plus what was added in the quench pot, is recycled to the EDC cleanup column. The hydrogen chloride, which would otherwise be a disposal problem, is pumped to the oxychlorination reactor, as shown in the upper right corner of Figure 9–2.

The oxychlorination reactor is packed with cupric (copper) chloride catalyst. Three feeds, gaseous hydrogen chloride, pure oxygen or oxygen in the form of air, and ethylene are reacted at 600–800°F and 60–100 psi, to form EDC, and water, as in Reaction Three in Figure 9-1. The reaction effluent is then piped over to the cleanup fractionator, where it commingles with the EDC stream from Reaction One and the recycle stream from VC fractionator.

Material Balance	
Feed:	
Ethylene	295 lbs.
Chlorine	750 lbs.
Product:	
Ethylene dichloride	1000 lbs.
By-products	45 lbs.
Feed:	
Ethylene dichloride	1667 lbs.
Product:	
Vinyl chloride	1000 lbs.
Hydrogen chloride	578 lbs.
By-products	89 lbs.

So, there are two recycle streams, hydrogen chloride and EDC. The EDC is recycled to pyrolysis; the hydrogen chloride is recycled to the oxychlorination to form EDC. Considerable attention has to be paid to balancing the flows around this plant. There are surge tanks in the plant that are not shown in Figure 9–2. But they can quickly fill up, potentially causing the need to shut down one of the reactions to play catch-up. Starting and stopping any of the reactions tends to be a problem, both in off-spec product and wasted energy costs.

Other technology

Chemists are always looking for ways to start with ethane and bypass ethylene on the way to the present ethylene derivatives. Now an old catalyst with a new twist has been developed, based on the work of Carl Ziegler and Giulio Natta. Chemists have immortalized these two researchers for the work they did in the mid-20th century by designating their contribution as the Ziegler-Natta class of catalysts.

Anyway, ethane can be converted to vinyl chloride monomer by passing it over a Ziegler-Natta type catalyst at 850–900° F in the presence of chlorine and oxygen. In a single vessel, oxychlorination of ethane to vinyl chloride monomer takes place:

$$CH_3\text{-}CH_3 \ + \ Cl_2 \ + \ {}^{1}\!/_2\,O_2 \ \rightarrow \ CH_2{=}CHCl \ + \ HCl \ + \ H_2O$$

| Ethane | Chlorine | Oxygen | Vinyl chloride | Hydrogen chloride | Water |

The process achieves about 90% conversion of ethane to VC. With the elimination of so many intermediate steps compared to the traditional EDC route, this process could achieve VC production cost savings of up to 35% anywhere an adequate supply of ethane can be found. That could even include the recycle stream from a heavy liquids olefins plant. If these killer economics persevere, this technology could grab all the growth in VC capacity and even replace most of the conventional VC capacity in a couple of decades. That's what happened to the acetylene-based route to VC when the ethylene-based route came on stream in the mid-20th century.

HANDLING CHARACTERISTICS

Sufficient evidence has proven that VC can cause cancer of the liver after prolonged exposure to only minute quantities (parts per million). Elaborate hardware precautions are taken to eliminate escape of any VC to the atmosphere. Personnel involved in production or use of VC often wear respirators whenever there is a possibility of a leak.

VC vaporizes at about 7°F, so at normal temperatures it must be contained in pressure vessels to keep it liquid. This includes movement by tank cars and trucks, which must fly the hazardous material sticker enroute.

Properties

EDC:

Molecular weight	98.96
Freezing point	-31.7°F (-35.4°C)
Boiling point	182.3°F (83.5°C)
Specific gravity	1.253 (heavier than water)
Weight per gallon	10.5 lbs/gallon

VC:

Molecular weight	62.5
Freezing point	-244.8°F (-153.8°C)
Boiling point	7.9°F (-13.37°C)
Specific gravity	0.9106 (lighter than water)
Weight per gallon	8.14 lbs/gallon

VC is highly reactive, and like styrene, will start to polymerize with itself if it just sits in a tank. Phenol, in trace amounts, is an effective polymerization inhibitor and is normally added to VC on the way to storage.

EDC is a much less nasty commodity. It need not be shipped in a pressurized vessel, but it is classified as a hazardous material and must be kept in a closed system.

COMMERCIAL ASPECTS

About 98% of EDC is used to manufacture VC. Relatively small amounts are used in the manufacture of perchloroethylene (an industrial degreaser and dry cleaning agent), in the manufacture of methyl chloroform (an anesthesia), and ethylenediamine (a fungicide and antifreeze inhibitor). About 99% of VC is used to manufacture polyvinyl chloride in a polymerization process described about 10 chapters from here. The rest of the VC goes into manufacturing chlorinated solvents.

EDC and VC are each traded commercially as a 99% pure grade. VC is usually designated as inhibited, indicating the presence of phenol to prevent spontaneous polymerization.

Chapter 9 in a nutshell...

Ethylene dichloride, $C_2H_4Cl_2$, is a petrochemical that is made so that vinyl chloride can be made out of it. The process for making EDC is sometimes integrated with the vinyl chloride plant. EDC is made by reacting ethylene with hydrogen chloride.

Vinyl chloride, C_2H_3Cl, or $CH_2=CHCl$, is ethylene with a chlorine atom attached, replacing a hydrogen. It is made in two ways. EDC is subjected to high temperatures in a cracking furnace, where a chlorine and a hydrogen atom pop off, leaving vinyl chloride. The availability of that free chlorine atom makes it appropriate to make hydrogen chloride, giving rise to the other route, where the hydrogen chloride, ethylene, and oxygen are reacted to make vinyl chloride and water. A single-step ethane-to-vinyl chloride process has commercial potential.

Almost all vinyl chloride is used to make polyvinyl chloride, a versatile consumer plastic.

EXERCISES

1. Mix and match:

quench	Addition of Cl_2 and O_2
oxychlorination	hot VC/HCl plus cool EDC
pyrolysis	C_2H_4 and Cl_2 to EDC
chlorination	EDC to VC and HCl

2. Take the three reactions for the EDC/VC plant and "net them out" algebraically, showing how much goes in the plant, how much comes out. (Hint: The only products are VC and water.)

3. What's the difference between chlorination, oxidation, and oxy-chlorination?

CHAPTER 10 ⬡

Ethylene Oxide and Ethylene Glycol

"From out of the past come the thundering hoofs
of the giant horse Silver."

**From *The Lone Ranger Rides Again*
Fran Striker**

In the preceding chapters, all the petrochemicals discussed, and their immediate derivatives had double-bonded carbons imbedded in their structure. This characteristic makes those chemicals very reactive, which is why they are so useful as building blocks. In contrast, ethylene oxide (EO) has no double bonds, but instead a three-member heterocyclic ring, with "*hetero*" meaning one of the atoms isn't carbon—it's oxygen. (*See* Figure 10–1.) This cyclic oxide is often called an epoxide. The suffix *ep-* is from the Latin meaning "on" or "beside." In chemistry, *ep-* generally refers to the heterocyclic ring. The other common chemical with this suffix is epichlorohydrin.

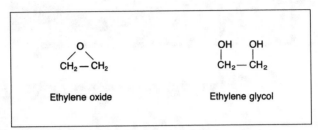

Fig. 10–1 Ethylene oxide and ethylene glycol

When EO is formed, single bonds from two adjacent carbons are connected to an oxygen atom. A three-member ring is always in a "strained" condition, due to the geometry of the molecule. Because of the propensity to relieve the strain, epoxides are very reactive. Practically all the EO produced is converted to chemical intermediates as a result of a ring opening reaction.

The key feature of ethylene glycol (EG) is the hydroxyl group, -OH, one on each of the two carbon atoms. The hydroxyls are responsible for its reactivity: EG is a monomer used in the production of polyester polymers. The hydroxyls also give EG its most important physical property: its solubility in water. That, linked with its low freeze point, makes EG suitable as an antifreeze and as a deicer. When EG is sprayed on ice, it combines with the water crystals and lowers the freeze point. This causes the mixture to melt and effectively keeps it in the liquid state.

ETHYLENE OXIDE

Until the 1940s, the commercial route to EO was ethylene

$$OH \quad Cl$$
$$| \qquad |$$

chlorohydrin, CH_2- CH_2. This was a two-step process—conversion of ethylene to the chlorohydrin by reaction with hypochlorous acid, HO-Cl, followed by dehydrochlorination (removal of HCl) of the ethylene chlorohydrin to give EO.

The problem with the process was not the yield of EO but the operating expenses, particularly the cost of chlorine. Almost all the chlorine introduced as part of the HO-Cl ended up after the process as calcium chloride. Not only was this compound a worthless solid, it created big time disposal problems.

In the 1940s and 1950s, a considerable amount of research was funded to find and develop the chemist's impossible dream: a process for the direct oxidation of ethylene to EO, without any by-products. Finally, Union Carbide found the silver bullet that did the job—a catalyst made of silver oxide. Silver oxide is the only substance found having sufficient activity and selectivity. (Activity relates to the amount of conversion, selectivity relates to the right yield.) Moreover, ethylene is the only olefin affected in this way. The others, propylene, butylene, etc., tend to oxidize completely, forming carbon dioxide and water. But when silver oxide is used as a catalyst with ethylene, the dominant reaction is the formation of EO. Some ethylene still ends up being further oxidized, as much as 25% in some processes, as shown in Figure 10–2.

The process was commercially so superior to the chlorohydrin route, that by the 1970s, the new chemistry had completely replaced the old. Adding some momentum to this transition was the fact that the obsolete and abandoned chlorohydrin plants could be readily converted to propylene oxide plants. The silver bullet for that process has yet to be found.

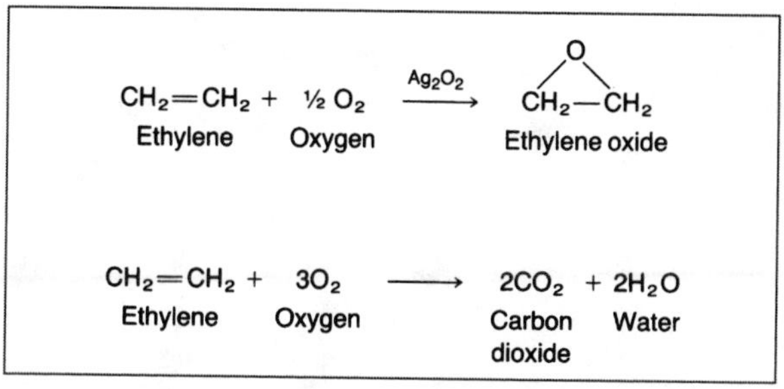

$$CH_2{=}CH_2 + \tfrac{1}{2}\,O_2 \xrightarrow{Ag_2O_2} \underset{\text{Ethylene oxide}}{CH_2{-}CH_2\ (O)}$$

Ethylene   Oxygen

$$CH_2{=}CH_2 + 3\,O_2 \longrightarrow 2CO_2 + 2H_2O$$

Ethylene   Oxygen   Carbon dioxide   Water

Fig. 10–2 Direct oxidation of ethylene to ethylene oxide

The process and the hardware

The new EO plants are as simple as any you will read about in this book. The feeds are mixed, reacted, then split into recycle and finished product streams, as shown in Figure 10–3. The oxidation reaction takes place in the vapor phase.

Compressed oxygen, and fresh and recycled ethylene, are heated, mixed, and then passed through a reactor with fixed beds of catalyst—silver oxide deposited on alumina pellets. In recent years the catalyst has been improved by the addition of promoters and inhibitors. (Promoters—in this case compounds of alkali or alkaline rare earth metals—enhance the activity of the catalyst; inhibitors—in this case chlorine compounds—chloroethane, or vinyl chloride, reduce its rate of activity decline.)

Like most oxidations, this one is exothermic. The temperature of the oxidation is controlled by the heat exchanger tubes built into the reactor. Water runs through the tubes, absorbs the heat of reaction, and turns to steam and exits the top. This keeps the reaction temperature at 500–550°F under slight pressure. The residence time of the feed in the reactor is only about one second. Yields, the amount of the ethylene that ends up as EO, approach 90%.

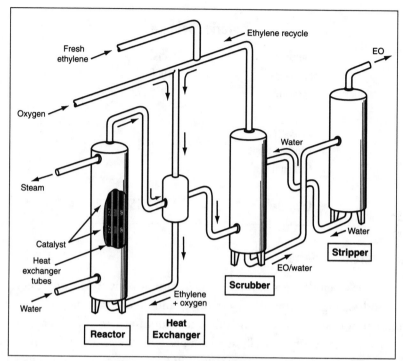

Fig. 10–3 EO plant

The effluent from the reactor is cooled in a heat exchanger. The EO, by-products, and unreacted ethylene are separated in a water-wash column in a manner just like the solvent recovery process described in Chapter 2. The EO is absorbed by the water while the by-products (mainly CO_2, plus the ever-present cats and dogs in small quantities), and unreacted ethylene are not. The EO/water solution is then steam-stripped and purified by fractionation.

The by-products and the ethylene are split, and the ethylene is recycled to the reactor. (The by-product splitter is not drawn correctly in Figure 10–3, because some of the by-products actually have lower boiling points than ethylene. The by-product splitter should really be shown as a series of columns.)

This process gives about 1.4 pounds of EO per pound of ethylene feed with a yield of 89%. If that seems confusing, see the definition of yield in the appendix.

Material Balance	
Feed:	**lbs.**
Ethylene	1000
Excess air	11,909 (2382 of O_2)
Product:	
Ethylene oxide	1,400
Carbon dioxide	343
Water	140
Unreacted air	11027 (1500 of O_2)

Commercial aspects

EO is a colorless gas at room temperature. It boils at 56°F. As a liquid it is colorless, highly flammable, explosive, and is very soluble in water and common solvents. EO is a toxic substance requiring care in handling.

EO is traded commercially as a high purity technical grade, 99.7% purity. Because of its low boiling temperature, EO must be stored and shipped in vessels that can withstand mild pressures. Trucks and tank cars must fly the red hazardous material label.

Properties	
Molecular weight	44.05
Freezing point	-169.0°F (-112.0°C)
Boiling point	56.3°F (13.5°C)
Specific gravity	0.8969 (lighter than water)
Weight per gallon	7.45 lbs/gallon

EO is an intermediary chemical and producers use it to make a variety of other things. The predominant derivative is ethylene glycol or EG (sometimes called monoethylene glycol or MEG, to distinguish it from diethylene glycol or DEG, and triethylene glycol or TEG), which uses up more than 60% of the total EO. EO is also used to make ethoxylates for further use in biodegradable detergents and as solvents for paints, lacquers, and other applications in the textile industry. Production of ethanolamine is the third major application for EO. It is also an excellent fumigant and sterilizing agent, which are the oldest uses for this commodity.

ETHYLENE GLYCOL

The conversion of ethylene oxide to ethylene glycol is one of the simplest processes in this book. The reaction, shown in Figure 10–4 requires that the EO ring be opened up, and hydroxyl groups formed. The reaction takes place in water with a little heat, pressure, and acid catalyst to promote it.

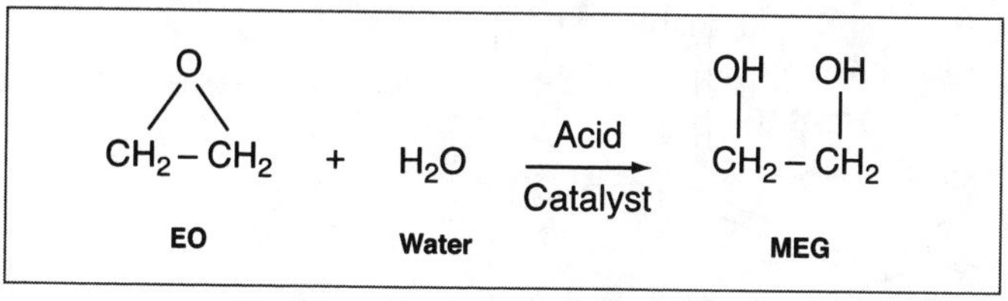

Fig. 10–4 EO to MEG Reaction

In the process shown in Figure 10–5, purified EO or a water/EO mixture is combined with recycled water and preheated to 400°F before being fed to a reactor. At a pressure of 200–300 psi, essentially all the EO is converted to MEG plus minor amounts of DEG and TEG. Water is kept in excess—generally 20 times as much water as the reaction would indicate—to assure high MEG selectivity. (Selectivity: the degree to which the targeted product is produced rather than other by-products.) Yield is essentially quantitative, i.e., the yield follows the amount indicated in the chemical equation in Figure 10–4. (*See* the Appendix for a clearer understanding of the word yield.)

The excess water is removed in a stripper column. The other three pieces of hardware in Figure 10–5—vacuum columns—are used to recycle the EO and to clean up the EG by splitting out the heavier glycol by-products.

High purity glycols result from this process, easily meeting the tight specifications set by polyester fiber and PET producers.

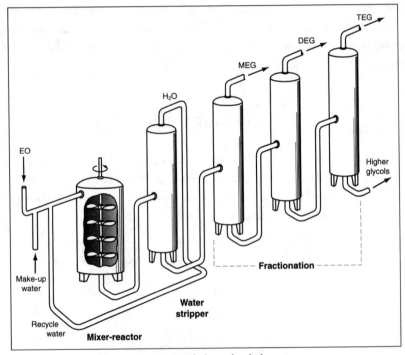

Fig. 10–5 Ethylene glycol plant

Material Balance

Feed:	lbs.
Ethylene oxide	721
Water (excess)	590
Product:	
Ethylene glycol	1000
Heavy glycols*	35
Unreacted water	5605

*di-, tri-, and tetraethylene glycols.

Since more than 60% of the EO production is converted directly to EG, the obvious question some macho chemist might ask is "why don't we do an end run and just convert ethylene directly to EG? Skip the oxidation step." Research starting 50 years ago led to several promising commercial processes, oxychlorination and acetoxylation. Exotic catalysts were used, and both avoided the EO step. But neither process was quite effective enough to replace the ethylene-to-EO-to-MEG route, which predominates today.

Commercial aspects

End uses. It's a little curious that the two major end uses for EG are so different. One is a consumer product; the other is a feedstock for more complicated chemistry. The reasons have to do with two separate properties of EG, one physical property, one chemical property. Because of EG's low freezing point, it is the main ingredient in automotive antifreeze. Because it is so chemically reactive, it is used as a monomer in making polyester polymers and PET, the plastic in the ubiquitous water and drink bottles.

Together, antifreeze, PET, and polyester polymers account for about 98% of the ethylene glycol produced in the United States. It is also used sometimes as a deicer for aircraft surfaces. The two hydroxyl groups in the EG molecule also make EG suitable for the manufacture of surfactants and in latex paints. Other applications include hydraulic brake fluid, the manufacture of alkyd resins for surface coatings, and stabilizers for water dispersions of urea-formaldehyde and melamine-formaldehyde. The hygroscopic properties (absorbs moisture from the air) make EG useful as a humectant for textile fibers, paper, leather, and adhesives treatment.

DEG and TEG are used as solvents for cellulose acetate derivatives and dyestuffs and as drying agents for refinery gases.

Ethylene Glycol Properties	
Molecular weight	62.07
Freezing point	11.3°F (-11.5°C)
Boiling point	387.7°F (197.6°C)
Specific gravity	1.1108 (heavier than water)
Weight per gallon	9.3 lbs./gal.

Properties and handling. Ethylene glycol is a clear, colorless syrupy, and virtually odorless liquid. It is hydroscopic, i.e., it absorbs water readily. And, when added to water, it lowers the freeze point. It is traded commercially as a high purity technical grade at 99% content. Ethylene glycol is a friendly liquid, and no particular precautions need to be taken in transporting it by barge, tank car, or truck, as long as you don't fall in it and lower your freeze point.

Chapter 10 in a nutshell ...

Ethylene oxide is a triangle-shaped cyclic compound. The "tightness" or shape of the bonds connecting the oxygen and the two CH_2 groups give ruse to EO's chemical reactivity. It readily converts to ethylene glycol in the presence of water at elevated temperatures. It also is used to make polymers and germicides. EO is made by direct oxidation of ethylene, which is greatly facilitated by the unusually effective catalytic power of silver oxide.

Ethylene glycol, CH_2OHCH_2OH, looks like ethane with hydroxyl groups (-OH) on each carbon in place of a hydrogen. EG is used as the essential ingredient in antifreeze and in the production of PET and polyester film, fiber, and plastics.

EXERCISES

1. Mix and match:

epoxide	EG
silver bullet	heavy glycol
epoxide ring	EO
antifreeze	CO_2
EO by-product	silver oxide, the EO catalyst
EG by-product	a molecule under stress

2. What caused the demise of the chlorohydrin route to EO?

3. The Clondike Coolant Company expects to sell 10 million gallons of ethylene glycol to antifreeze blenders this summer. They'll add some additives (maybe), some blue or red or yellow dye, and maybe some water and mark it up about 400%. How much ethylene should Clondike contract for, theoretically, to feed to their ethylene oxide-ethylene glycol plants this spring?

CHAPTER 11

Propylene Oxide and Propylene Glycol

"Something old, something new,
something borrowed . . . "

Traditional wedding rhyme

You have to talk about propylene oxide and propylene glycol after ethylene oxide and glycol. It's not that the chemical configurations are so similar (they are), or that the process chemistry is about the same (it is). The fact is that much of the propylene oxide is now made in plants originally designed and constructed to produce EO, not PO. As you read in the last chapter, the chlorohydrin route to EO was abandoned by the 1970s in favor of direct oxidation. At the same time, the EO producers found that the old EO plants were suitable for the production of PO and certainly the cheapest hardware available to satisfy growing PO demands.

PROPYLENE OXIDE

The chemical structure of PO differs from EO by the methyl group ($-CH_3$), as shown in Figure 11–1. The difference is more than just a matter of geometric symmetry. The methyl group in PO, for example, increases the reactivity of the molecule in an adverse way. In the various manufacturing routes to PO, consequently, the reaction of propylene with chlorine or oxygen is hard to stop. By-products are an unavoidable nuisance, and the yields of PO are not as high as the chemical engineers would like.

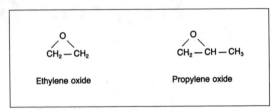

Fig. 11–1 Ethylene oxide and propylene oxide

Prior to the late 1970s, almost all PO was produced using the chlorohydrin route, much of it in the former EO plants. But this process was encumbered with the same problems as the EO—it was energy intensive, byproduct yield was too high, and the chlorine waste product was disposal expense. As a consequence, newer technology emerged in the 1980s—the indirect oxidation route. It involves the oxidation of a hydrocarbon (call it R-H) to form a hydroperoxide (with the signature, R-OOH) that is then reacted with propylene to form PO and an alcohol coproduct (having the signature R-OH). The market value of the alcohol assists materially in justifying the economics of this alternate route.

The commercial success of the indirect oxidation route has dampened enthusiasm for continuing the search for the "silver bullet" catalyst that will facilitate direct oxidation, *a la* EO. Further activity in that area seems more and more like the quest for that Philosopher's Stone that turns lead into gold.

The chlorohydrin route

The chlorohydrin route takes two steps. The description of them, unfortunately, is a sentence whose average word length is nine letters: reaction of propylene with hypochlorous acid (HO-Cl) followed by dehydrochlorination of the propylene chlorohydrin with calcium hydroxide. That's a tough way of saying that a chlorine atom (Cl-) and a hydroxyl group (-OH) are added to the propylene double bond, and then a chlorine atom and a hydrogen atom are removed, leaving the oxygen bonded to two adjacent carbon atoms to form propylene oxide. (*See* Figure 11–2.)

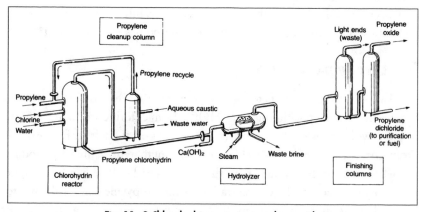

Fig. 11–2 Chlorohydrin route to propylene oxide

Three equations describe the process. The first involves making the hypochlorous acid by reacting chlorine and water. In the second, the acid reacts with propylene to make chlorohydrin. The dehydrochlorination takes place in the third to give propylene oxide.

Hypochlorous Acid
$$Cl_2 + H_2O \rightarrow HOCl + HCl$$

Formation of Propylene Chlorohydrin

$$CH_3-CH=CH_2 + HOCl \rightarrow CH_3-\underset{\underset{OH}{|}}{CH}-\underset{\underset{Cl}{|}}{CH_2}$$

Dehydrochlorination to PO

$$2CH_3-\underset{\underset{OH}{|}}{CH}-\underset{\underset{Cl}{|}}{CH_2} + Ca(OH)_2 \rightarrow CH_3-\overset{\overset{O}{/\ \backslash}}{CH-CH_2} + CaCl_2 + 2H_2O$$

One other reaction not shown is the formation of propylene dichloride. The demand for this compound is generally insufficient to absorb all the coproduction, so it also ends up on the list of "things to be disposed of coming from the PO-chlorohydrin process." But despite this and all the other problems already mentioned about the chlorohydrin route, the process remains economically healthy—breathing heavily, but healthy. Indeed, 40 to 50% of the PO produced in the United States comes from this route.

The chlorohydrin hardware

Two of the reactions take place in the same reactor in this plant. The formation of the hypochlorous acid (HOCl) from chlorine and water, and the reaction with propylene all occur simultaneously on the left in Figure 11–2. Propylene reacts readily with chlorine to form that unwanted by-product, propylene dichloride. To limit that, the HOCl and HCl are kept very dilute. But as a consequence, the concentration of the propylene leaving the reactor is very low—only 3–5%! At any higher concentration, a separate phase or second layer in the reactor would form. It would preferentially suck up (dissolve) the propylene and chlorine coming in, leading to runaway dichloride yields. The low concentration levels of the propylene chlorohydrin and the need to recycle so many pounds of material is the reason the process is so energy intensive. It just takes a lot of electricity to pump all that stuff around.

The unreacted propylene is taken off the top of the reactor and cleaned up for recycling. By bubbling this stream through a dilute caustic solution (like sodium hydroxide, NaOH), the chlorine and HCl carried along with the propylene are removed by converting them to sodium chloride, NaCl, and water. The "scrubbed" propylene is then taken overhead (from the top of the fractionation column) and is ready as fresh feed or use elsewhere in the plant.

The dilute propylene chlorohydrin stream is mixed with a solution of water and 10% slaked lime, calcium hydroxide, and pumped to a vessel called the hydrolyzer. The chlorohydrin rapidly dehydrates to PO. The reaction is so fast that the PO has to be sprung from the mixture before the reaction continues, forming propylene glycol. Steam is bubbled through the reactor, helping to flash (vaporize) the PO out of the reaction zone.

The vapor from the hydrolyzer contains not only water and PO, but also propylene dichloride and whatever other cats and dogs (by-products). Fractionation columns are used to purify the PO to a 99% technical grade.

Material Balance
(PO via the chlorohydrin route)

Feed:	lbs.
Propylene	941 lbs.
Chlorine	1590 lbs.
Slaked line (CA(OH))$_2$	636 lbs.

Product:	
Propylene oxide	1000 lbs.
Calcium chloride	955 lbs.
Hydrogen chloride	628 lbs.
Propylene dichloride	437 lbs.
By-products	147 lbs.

The indirect oxidation route

This indirect oxidation route takes two steps. In the first, a hydrocarbon, such as isobutane or ethylbenzene, is oxidized. The source of the oxygen is air. The reaction takes place just by mixing the ingredients and heating them to 250–300°F at 50 psi, producing a hydroperoxide. In the second step, the oxidized hydrocarbon reacts with propylene in a liquid phase and in the presence of a metal catalyst at 175–225°F and 550 psi to produce PO yields of better than 90%. The process flow is shown in Figure 11–3.

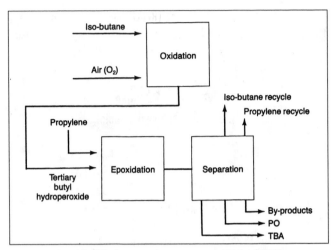

Fig. 11–3 Indirect oxidation route to PO (and TBA)

The reaction sequence is summarized below using isobutane as the hydrocarbon. The crucial and nearly incredible part of the process is in two parts itself but only shows as one. It is the second equation where the oxygen molecule transfers to the propylene molecule and the ring closes to form the epoxide. (That's why they call it epoxidation). The magic that causes all that to happen is in the metal catalysts, molybdenum naphthenate or the soluble salts of titanium, vanadium, or tungsten. This molecular fancy-dance is but one of many examples in chemistry where catalysts can cause atoms to slide around molecules in unlikely ways.

Formation of Hydroperoxide (Oxidation of Isobutane)

$$
\begin{array}{ccc}
 & & \text{OOH} \\
 & & | \\
\text{CH}_3\text{–CH–CH}_3 + \text{O}_2 & \rightarrow & \text{CH}_3\text{–C–CH}_3 \\
| & & | \\
\text{CH}_3 & & \text{CH}_3 \\
\text{Isobutane} & \text{Oxygen} & \text{Tertiary butyl hydroperoxide} \\
 & & \text{(TBH)}
\end{array}
$$

Epoxidation of Propylene (Reaction with Propylene)

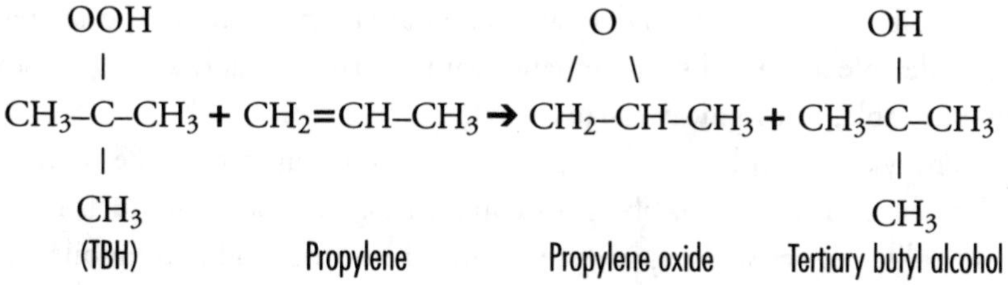

$$
\begin{array}{c}
\text{OOH} \\
| \\
\text{CH}_3\text{–C–CH}_3 \\
| \\
\text{CH}_3 \\
\text{(TBH)}
\end{array}
+ \;\text{CH}_2\text{=CH–CH}_3 \;\rightarrow\;
\overset{\displaystyle \text{O}}{\overset{\displaystyle /\;\backslash}{\text{CH}_2\text{–CH–CH}_3}}
+ \;
\begin{array}{c}
\text{OH} \\
| \\
\text{CH}_3\text{–C–CH}_3 \\
| \\
\text{CH}_3
\end{array}
$$

Propylene   Propylene oxide   Tertiary butyl alcohol

One of the commercial benefits of this route is the value of the coproducts, tertiary butyl alcohol (TBA) when isobutane is used, and styrene when ethylbenzene is used. TBA also can be easily hydro-treated back to isobutane if a recycle stream for PO manufacture is more advantageous.

Material Balance
(PO via the Indirect Oxidation Route)

Feed:

Propylene	782 lbs.
Isobutane	1880 lbs.
Oxygen (in excess)	1069 lbs.
Catalyst	small

Product:

Propylene oxide	1000 lbs.
Tertiary butyl alcohol	2400 lbs.
Unreacted oxygen	269 lbs.
By-products	62 lbs.

Commercial aspects

Although propylene oxide is structurally similar to ethylene oxide, its applications are very different. For example, propylene glycol accounts for only about 20% of PO demand, compared to the 60–65% share EG takes of the EO. About 60% of the PO is used to make polyethers and polyether polyols. These are chemical derivatives that are reacted with diisocyanates to make flexible and rigid polyurethane foam. You're probably sitting on some now (flexible, not rigid, if you're lucky).

Propylene glycols and dipropylene glycols are used to make thermoset polyester resins for use in fiberglass composites. Fabricated products include boat hulls, shower stalls, appliance casings, furniture, and automobile parts. The polyester is usually reinforced with shredded, chopped, or woven glass fiber. Ironically, even though boats are said to have fiberglass hulls, most of the material by weight is polyester made from propylene oxide.

Smaller but growing markets for PO include liquid detergents and surface coatings. The PO route to butanediol is replacing the acetylene route.

Propylene Oxide Properties	
Molecular weight	58.08
Freezing point	-155.2°F (104.0°C)
Boiling point	93.6°F (34.2°C)
Specific gravity	0.826 (heavier than water)
Weight per gallon	6.92 lbs./gal.

Propylene oxide is a low boiling point, flammable liquid, readily soluble in both water and the more common organic solvents, such as alcohol, ether, and aliphatic and aromatic hydrocarbons. Commercial sales involve only technical grade (about 98%), and bulk movements require a hazardous material shipping label. Standard transport equipment (trucks, tank cars, and barges) can be used.

Propylene Glycol

This has to be the quickest product treatment in this book—if you've already read the ethylene glycol chapter. The process for propylene glycol is the same as for EG. A little sulfuric acid in water at about 150°F will open up the epoxide ring, and the water will provide the hydroxyl groups to form propylene glycol. With plenty of excess water, high yields of propylene glycol are achieved. However, some higher glycols, primarily dipropylene glycol, will show up as by-products.

$$
\underset{\begin{array}{c}/ \ \backslash\end{array}}{\overset{O}{CH_3-CH-CH_2}} + H_2O \rightarrow \underset{\begin{array}{c}| \quad |\end{array}}{\overset{OH \ OH}{CH_3-CH-CH_2}}
$$

(Glycol is from the Greek route, *glyk-*, meaning sweet. The link is through the sugars, which have structures much like propylene glycol, with multiple carbons and hydroxyl groups.)

The hardware for propylene glycol is the same as that shown in Figure 10–5. Just substitute propylene for ethylene to identify the streams.

Material Balance

Feed:

Propylene oxide	887 lbs.
Water	275 lbs.
Catalyst	Trace

Product:

Propylene glycol	1000 lbs.
Dipropylene glycol	130 lbs.
Tri-, tetra-, and heavier glycols	32 lbs.

Commercial aspects

Uses. Polyester resins use up about 60% of the propylene glycol (and most of the dipropylene glycol) manufactured. The remainder is used as a tobacco and cosmetic humectant (a chemical that keeps moisture around), automotive antifreeze and brake fluid ingredients, food additive, and plasticizers for various resins, and making nonionic detergents and coatings. Propylene glycol is an excellent solvent.

Many of the derivatives of propylene glycol, namely the ethers and the acetates, behave very much like the corresponding ethylene glycol derivatives. For that reason they are easily substitutable for each other.

Propylene Glycol Properties

Molecular weight	76.11
Freezing point	76.0°F (-60.0°C)
Boiling point	361.1°F (183.7°C)
Specific gravity	1.0381 (heavier than water)
Weight per gallon	8.72 lbs./gal.

Properties and handling. You can tell from the applications that propylene glycol is safe. It is nontoxic, nonflammable, and even fit for human consumption (in small doses). It is a colorless, odorless, sweet-tasting liquid, completely miscible or soluble in water. Propylene glycol is available in three grades: NF (99.99%), technical (99%), and industrial (95%).

Chapter 11 in a nutshell ...

Propylene oxide is a triangle-shaped cyclic compound with a -CH$_3$ group "along the bottom." An oxygen atom bonded to two adjacent carbon atoms makes the triangle. No catalyst has been found to efficiently oxidize propylene directly, so the commercial processes involve two steps. The chlorohydrin route uses propylene, chlorine, and oxygen. The indirect oxidation route involves isobutane, air, and propylene. The latter also produces tertiary butyl alcohol, a valuable gasoline-blending component. Ethylbenzene can be substituted for isobutane. In this case, styrene is coproduced with the propylene oxide.

Propylene oxide is a liquid at room temperature. It readily converts to propylene glycol in the presence of water and a little acid. But the majority of propylene oxide is used to make polymers, including polyurethane foam.

Propylene glycol looks like propane with hydroxyl groups, -OH, substituted for a hydrogen on two adjacent carbons. Propylene glycol is also used to make polymers and in a variety of smaller applications such as solvents, humectants, and food additives.

EXERCISES

1. Fill in the blanks:

    a. Forming a ringed compound like propylene oxide from propylene
        is called _____.
    b. Two commercial routes to propylene oxide are _____ and
        _____. The _____ _____ route
        has not been commercially successful.
    c. Commercially valuable by-products of propylene oxide and of
        propylene glycol manufacture mentioned in this chapter are
        _____, _____, and _____.

2. Why is the chlorohydrin route losing favor as the preferred route to
    propylene oxide:

3. Write the equation for the indirect oxidation of ethylbenzene to propy-
    lene oxide and styrene.

COMMENTARY 2 ⬡

"Is you deaf?" sez Brer Rabbit sezee.
"Kaze if you is, I can holler louder," sezee.

The Wonderful Tar-Baby Story
Joel Chandler Harris, 1848–1908

REVIEW

You can consider the Mutt and Jeff/Gemini chemicals introduced in the last five chapters the first and second derivatives of the petrochemical building blocks ethylene, propylene, and benzene, as shown in the following figure. The first line derivatives are all catalyst induced. Pressure and temperature are the operatives. Styrene and vinyl chloride are a little special. The characteristically reactive double bond, in both of them, is induced by high-temperature pyrolysis cracking off the required hydrogen atoms. That makes these second level derivatives behave much like the basic building blocks ethylene and propylene. In fact, their major applications are in polymers, similar to much of the two olefins' volume.

There is another route to propylene oxide and to styrene not shown on the figure. Indirect oxidation of isobutane or of ethylbenzene just doesn't fit neatly on the chart, but both are commercial processes, the former yielding PO and TBA, the latter giving PO and styrene.

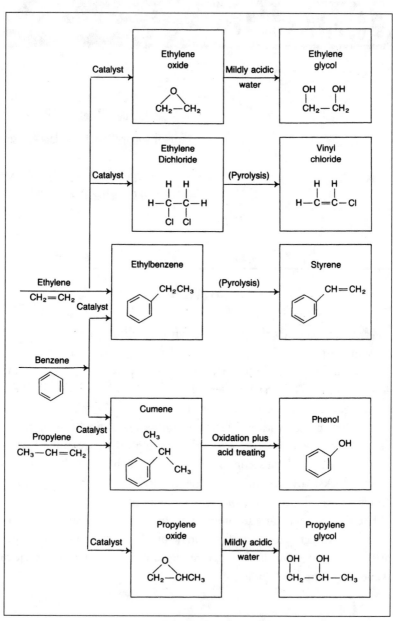

Primary and secondary derivatives

FOREWORD

The next 10 chapters cover a collection of petrochemicals not altogether related to each other. Synthesis gas is a basic building block that leads to the manufacture of ammonia and methanol. MTBE is made from methanol from synthesis gas (with a little isobutylene thrown in). The alcohols in Chapter 14 and 15, the aldehydes in 16, the ketones in 17, and the acids in 18 are all closely related to each other by looks, though the routes to get to them are perplexingly different. Alpha olefins and the plasticizer and detergent alcohols have the same roots and routes, but different ones from the rest. Maleic anhydride, acrylonitrile, and the acrylates—well, they're all used to make polymers and they had to be somewhere.

CHAPTER 12 ◇

METHANOL AND SYNTHESIS GAS

"As the poet said:
"Only God can make a tree"—
probably because it's so hard to get the bark on.

Without Feathers
Woody Allen, 1935–

The order in which you have to approach these two subjects is reversed in the chapter title because you might be apprehensive if they weren't. After all, everyone knows what methanol is. It's methyl alcohol, CH_3OH, wood alcohol, carbinol, or if you're a student of medieval culture, aqua vitae. But what is "synthesis gas?" It's not a familiar name because it's not usually handled in commercial transactions. The term synthesis gas refers to various mixtures of carbon monoxide (CO) and hydrogen (H_2) used for the manufacture of certain petrochemicals. In the 19th century, it was produced by passing steam over coke at very high temperatures. Today it's made largely from natural gas (methane). But a few paragraphs about synthesis gas, how it's made, and how it can be used in the synthesis of other petrochemicals will be beneficial. That's particularly true because two important chemicals, ammonia and methanol, are derived from synthesis gas.

SYNTHESIS GAS

Nature hasn't provided any convenient sources of pure CO and H_2. There's some of each contained in natural gas but usually not in sufficient quantities to justify going after it. But these two compounds, either in the combined state or separate, are readily convertible to a number of commercial compounds. With that as a motivator, several processes have been developed to convert natural gas to synthesis gas. Natural gas is largely methane (CH_4), and that provides a source of carbon and hydrogen. Air or water provides the other necessary ingredient, oxygen.

Synthesis gas can easily be confused with the oxymoron synthetic natural gas, SNG. Both are sometimes called "syngas." But SNG is basically methane made from petroleum products, like naphtha or propane, or from coal. It's used as a substitute for or supplement to natural gas.

The synthesis gas processes

The two predominant methods of making synthesis gas are steam reforming and partial oxidation. Both are quite simple. The steam reforming method involves passing methane or naphtha plus steam over a nickel catalyst. The reaction, if methane is the feedstock, is:

$$CH_4 + H_2O \rightarrow CO + 3H_2$$

The reaction relies on the brute force of high temperatures and pressures and must be carried out in hardware much like the cracking furnaces described in the ethylene chapter. As always with cracking, undesirable reactions occur, resulting in the formation of CO_2 and carbon. The latter is particularly a nuisance because it sets down on the catalyst and deactivates it.

The other method is the partial oxidation of methane:

$$CH_4 + \tfrac{1}{2}O_2 \rightarrow CO + 2H_2$$

Like the steam reforming method, this process takes place at severe conditions, high temperatures and pressures, but no catalyst. The reaction is called partial oxidation because it is kept from going to CO_2 by limiting the amount of oxygen fed to the process.

The partial oxidation method is normally used for heavier feedstocks, everything from naphtha to residual fuel, in those places where natural gas or light hydrocarbons (ethane, propane, or butane) are not readily available.

The yield of CO is not 100% in either process. You can see in Table 12–1 that plenty of CO_2 also gets formed as a by-product.

	H_2	CO	CO_2	TOTAL
Steam reforming	75	15	10	100
Partial oxidation	50	45	5	100

Table 12–1 Synthesis gas composition (percent yield based on methane feed)

Fortunately, CO_2 can be removed without too much difficulty by solvent extraction. Even better, it can then be reacted with steam and more methane to give off CO and H_2. This step is also done at high temperatures and pressures, and a nickel catalyst is used.

$$3CH_4 + 3CO_2 + 2H_2 \rightarrow 6CO + 8H_2$$

In addition, this step is sometimes used to supplement the other reactions to get the proper combination of CO and H_2, since the CO:H_2 ratio is so different. Making synthesis gas for methanol production, for example, needs this help.

Synthesis gas can be tailored in this manner to fit any number of specific applications. For example, a commercial route to aldehydes (the R-CHO signature group) and alcohols (the R-OH signature group) is the Oxo reaction, as discussed in the section on normal butyl alcohol in Chapter 14. In that reaction, the CO:H_2 ratio needed is 1:1. Careful adjustment of the three feedstocks, CH_4, CO_2, and H_2O and the amount of recycling will give this combination.

Commercial aspects

Most of the synthesis gas produced is captive. That is, it's consumed by the manufacturer. Synthesis gas plants are normally integrated into the adjacent application plant. When there is a two-party transaction involved, the properties of the synthesis gas stream are normally specified in a contract. There are no universally accepted standards that apply with this stream.

The only practical way to move synthesis gas around is by pipeline, and even in two-party transactions, the pipelines are usually no longer than a mile or two. Beyond that, the pipeline capital cost starts to affect the economics of the applications.

Ammonia

The most important uses of synthesis gas are the manufacture of ammonia (NH_3) via the Haber process. A mixture of nitrogen and hydrogen are passed over an iron catalyst (with aluminum oxide present as a "promoter"). The operating conditions are extreme—800°F and 4000 psi.

$$N_2 + 3H_2 \rightarrow 2NH_3$$

Why synthesis gas? And where does the nitrogen come from? Synthesis gas, of course, provides the hydrogen; air provides the nitrogen. And if the synthesis gas process is partial oxidation, then there was probably an air separation plant associated with it. That separates the oxygen from the nitrogen for making the synthesis gas, and leaves the nitrogen for feed to the ammonia plant.

In most ammonia plants, there are facilities to remove CO from the feed because CO will poison the catalyst. Generally, the technique used is to react the CO with water to produce CO_2 and H_2. The CO_2 is removed by solvent extraction, and the H_2 is recycled. (In case you were wondering, typical solvents used to remove CO_2 are ethanolamine or an aqueous solution of potassium carbonate.)

METHANOL

There's a good reason why methanol is commonly called wood alcohol. The early commercial source was the destructive distillation of the fresh-cut lumber from hardwood trees. When wood is heated without access to air at temperatures above 500°F, it decomposes into charcoal and a volatile fraction. Among the compounds in the volatile fraction is methanol. Hence, the name wood alcohol or wood spirits.

Since 1923, methanol has been made commercially from synthesis gas, the route that provides most of the methanol today. The plants are often found adjacent to or integrated with ammonia plants for several reasons. The technologies and hardware are similar, and the methanol plant can use the CO_2 made in the Haber ammonia process. In this case, the route to methanol is to react the CO_2 with methane and steam over a nickel catalyst to give additional CO and H_2 and then proceed to combine these to make methanol:

$$3CH_4 + 2H_2O + CO_2 \rightarrow 4CO + 8H_2$$
Synthesis gas

$$CO + 2H_2 \rightleftarrows CH_3OH$$
Methanol

The double arrows in the methanol reaction indicate that the reaction can go in either direction. There is a principle here that is taught in the sophomore P-chem class (Physical chemistry) of every chemical engineer. Methanol, in the vapor state, takes up only one-third the volume as the equivalent amounts of CO and H_2. So in order to "push the reaction to the right," the process is run under pressure. That causes the compound that takes up less volume to be favored—synthesis gas to methanol rather than methanol to synthesis.

The first commercial plant that converted synthesis gas to methanol was built in 1924 in Germany by BASF. It ran at very high pressures (3500–5000 psi) and used a zinc-copper catalyst. In the years since then, further development of catalysts has brought the pressures down, eliminating much of their expensive capital and operating costs. In the 1950s, medium pressures of 1500–3500 psi were in vogue. At the present, newer catalyst based on copper-zinc oxide have resulted in lower pressures—1000–1500 psi in 90% of the plants.

The process is still expensive, and that continues to give incentives to ongoing research to find the elusive catalyst that will permit direct conversion of methane to methanol without having to break apart the methane and reassemble it again. Breakthroughs in this technology are possible at any time, which could obsolete this whole sector of the petrochemical industry. While that's true of many parts of the industry, this process seems somewhat more vulnerable. Meanwhile, new single train plants with capacities of 5000 metric tons per day are putting economic pressure on older units with capacities at only 1000–2500.

The plant

The process for synthesis of methanol involves these basic steps:

1. Steam reforming of natural gas plus addition of CO_2 to adjust the $CO:H_2$ ratio to 1:2
2. Compression to 1000–1500 psi
3. Synthesis in a catalytic converter operating at 400–500°F
4. Purification—distillation

The hardware is shown in Figure 12–1. To protect the compressors, a water knockout column in front is necessary. It keeps water slugs from forming during compression, sending turbine blades flying all around the plant.

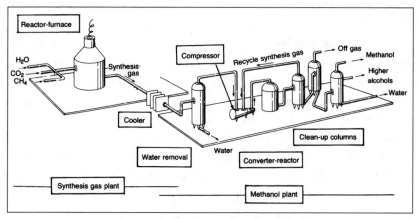

Fig. 12–1 Synthesis gas and methanol plants

The compressed gas is heated and passed through a reactor that has baskets of catalyst. In between the baskets are heat exchangers. The reaction is exothermic, but the reaction is sensitive to the temperature, so heat must be rapidly removed.

The effluent from the reactor contains only 5–20% methanol because the one-pass conversion is very low. After cooling and pressure letdown, the liquid methanol can be removed and further purified by distillation. The unreacted synthesis gas is compressed and recycled to the reactor.

Methanol of 99% purity is obtained. By-products are 1–2% dimethyl ether (CH_3OCH_3), about 0.5% higher alcohols (ethyl, propyl, isobutyl, and higher), and some water.

Material Balance

Feed:

Carbon monoxide	921 lbs.
Hydrogen	132 lbs.

Product:

Methanol	1000 lbs.
By-products	53 lbs.

Commercial aspects

Properties and handling. Methanol is a colorless, volatile liquid at room temperature with an alcoholic smell. It mixes with water in all proportions and burns with a pale blue flame. Methanol is highly toxic. As little as a fifth of a shot (10cc) can cause blindness. Larger amounts kill. It should never be applied to the body as a rubbing alcohol because the vapors are so toxic.

Sale grades of methanol include 95 and 97% purity. U.S. Federal grade must meet 99.8% minimum purity and be acetone-free. Methanol can be transported in conventional tank trucks, rail cars, ships, and barges, but it must be in closed systems. The red hazardous materials markings are required.

Methanol Properties	
Molecular weight	32.04
Freezing point	-143.7°F (-97.6°C)
Boiling point	148.3°F (64.6°C)
Specific gravity	0.792 (lighter than water)
Weight per gallon	6.59 lbs./gal.

Uses. About 35–40% of the methanol made is converted to formaldehyde. That's not because the embalming business is so good. Formaldehyde is a feedstock for amino and phenolic resins, which are used as adhesives in plywood, and in the automotive and appliance industry to make parts (all the agitators in washing machines used to be made out of phenolic resins). It is used as feedstock for hexamethylene tetramine, used in electronic plastics; for pentaerythritol, used for making enamel coatings and for floor polish and inks; for butanediol, a chemical intermediate; and for acetic acid, which is widely used itself as a feedstock and solvent and warrants its own treatment later on. In the textile business, formaldehyde is used to make fire retardants, mildew resistant linens, and permanent press clothing.

Another application of methanol is the production of methyl chloride, which is used in making silicone rubber, including the caulking and sealing compounds that will set at room temperature (the kind you buy in a tube at the hardware store.)

What has been the fastest growing use of methanol is not in the petro-chemical industry at all but in the automotive fuel business. As much as 40% of the methanol produced ends up in gasoline via two routes, the manufacture of methyl tertiary butyl ether (MTBE), a gasoline-blending compound, and as a direct substitute for gasoline—either in part as a gasohol blend or in total.

Atmospheric pollution problems have focused public policy on the elimination of the pollutants formed during the combustion of gasoline in highway vehicles. The formation of ozone, nitrogen oxides, and some sulfur oxides has been thought to be alleviated by the addition of organic compounds that contain oxygen in them, such as MTBE and methanol. Commercial testing of gasoline blends with these compounds and local laws requiring minimum content became prevalent in the United States in the late 1980s. Ironically, Baddour's Law[1] came into effect again when MTBE was found to be polluting groundwater through spillage and leaky under-ground tanks. Already banned in some parts of the United States, MTBE may ultimately be phased out of all gasoline.

Nevertheless, as motor fuels, MTBE and methanol have different appeals. MTBE has a high-octane number, which makes it suitable for blending premium gasoline. Methanol has a lower octane number but is cheaper to manufacture. Still it is generally considered more expensive than gasoline on an economic basis as long as the price of crude is under about $30 per barrel. A gallon of methanol, by the way, is not equivalent to a gallon of gasoline. The energy content of gasoline is about 1.8 times that of methanol, and the miles per gallon in an equivalent engine are in about the same proportion.

Methanol is also used to "de-nature" ethanol. There's not much differ-ence between synthetic ethyl alcohol and the "real thing" made from rye and other grain. Methanol is added, for political reasons, up to 10%, to the synthetic stuff to keep it from being substituted for the "real" or "natural" ethyl alcohol. Ethanol "de-natured" in this way is toxic enough to cause headaches, dizziness, vomiting, blindness, and coma, depending on how much is consumed. That's usually a sufficient threat.

Smaller volumes of methanol are used in the production of dimethyl terephthalate that goes into polyester fibers and in methyl methacrylate, which goes into plastics.

Chapter 12 in a nutshell ...

Synthesis gas is a loose name for hydrogen/carbon monoxide mixtures of varying proportions. These two compounds are so basic, they are too simple to start with for most petrochemicals. The primary applications of synthesis gas are only ammonia and methanol manufacture and a little normal butyl and 2-ethylhexyl alcohol production.

Synthesis gas is made by decomposing methane (from natural gas) in the presence of water. The reaction takes place at high pressure and temperature in the presence of a catalyst. The proportion of H_2 and CO depends on the amount of CO_2 that is left in the product stream or is recycled to be converted to CO/H_2. Most synthesis gas plants are built adjacent to the plants where the synthesis gas will be used.

Methanol, CH_3OH, the simplest alcohol, is made by reacting CO and H_2 at high pressures over a catalyst. Methanol is a liquid at room temperature and is highly toxic. It is used to make formaldehyde, acetic acid, and other chemical intermediates. It is also used as a feedstock for MTBE (methyl tertiary butyl ether), a gasoline-blending component.

EXERCISES

1. You need two H_2 molecules for each CO molecule to make methanol, CH_3OH. How do you get CO and H_2 in nearly the right proportions so you waste as little as possible of either?

2. What's the synergy or affinity of ammonia and methanol plants?

3. What are the largest applications of methanol?

Endnotes for Chapter 12

[1] Professor Robert F. Baddour, Professor of Chemistry at the University of Texas. once declared, "You cannot eliminate one pollutant without creating another."

CHAPTER 13 ⬡

Methyl Tertiary Butyl Ether

[He] throws aside his paint pots
and his words a foot and a half long.

Ars Poetica
Horace, 65–8 B.C.

I f it weren't for the social and political pressure put on the petroleum refining industry in the 1970s, methyl tertiary butyl ether (MTBE) would not have enough volume to qualify for a chapter in this book. Prior to that time, paint and varnish makers used almost all the MTBE made as a solvent; petrochemical companies used a small amount as a solvent for extraction, and that was it. But environmental restrictions forced gasoline producers to look first for a high-octane additive to replace the lead additives that were being mandated out of gasoline. MTBE has a Research Octane Number of 118. Then an even higher wave hit MTBE when governments mandated a minimum oxygen content in gasoline to enable more complete combustion and reduce emissions of volatile organic compounds.

The tides of fortune rise and ebb, and by the 21st century, MTBE was found to be polluting aquifers. In addition, the efficacy of adding oxygenates to gasoline as older cars were junked became contentious. The bloom fell off the MTBE rose, and the petroleum industry has to decide what to do with all the MTBE capacity, most of which is in refineries.

THE PROCESS

The ingredients for MTBE are isobutylene and methanol. Historically, refiners have noticed a perennial surplus of methanol-producing capacity around the world. Furthermore, huge surpluses of natural gas in remote locations promise even more methanol supply should the market price ever rise slightly above investment values. In a shrewd move, most refiners have opted to buy their methanol requirements rather than make them.

Isobutylene supply initially came mainly from the cracked gas streams generated by the cat cracker, plus whatever other units fed gases into the cracked gas plants. The isobutylene market trades very thinly, so when the cracked gas plant supply is insufficient, a producer must turn to a dehydrogenation process for converting isobutane to isobutylene.

The reaction uses equal molar parts of methanol and isobutylene:

$$CH_3OH + CH_3-\underset{\underset{\displaystyle CH_3}{|}}{\overset{\overset{\displaystyle CH_3}{|}}{C}}=CH_2 \rightarrow CH_3-\underset{\underset{\displaystyle CH_3}{|}}{\overset{\overset{\displaystyle CH_3}{|}}{C}}-O-CH_3$$

A number of companies offer their own proprietary process designs for making MTBE. Since MTBE came out of the petrochemicals industry, many of them have a more exotic process than refiners normally deal with, called *catalytic distillation.* This apparatus combines a distillation column with a reactor in a single vessel and is often used to complete a reaction started in another reactor, as shown in Figure 12-1. The column has a catalyst bed in the middle and trays above and below it. The idea is to introduce the feed into the catalyst bed. The catalyst causes a reaction that generates enough heat to cause any unreacted feed to vaporize while the, reaction product, MTBE, remains a liquid. The sets of trays at the top and the bottom of the column then assure a good clean separation of the unreacted reactants (if that makes sense) and the MTBE.

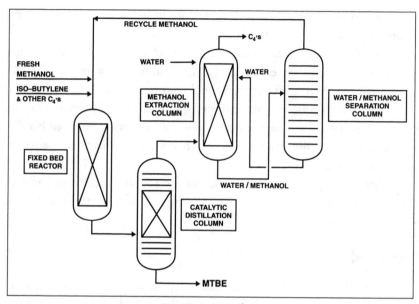

RECYCLE METHANOL

C₄'s

FRESH
METHANOL

WATER →

ISO–BUTYLENE
& OTHER C₄'s

WATER

WATER

METHANOL
EXTRACTION
COLUMN

WATER / METHANOL
SEPARATION
COLUMN

FIXED BED
REACTOR

WATER / METHANOL

CATALYTIC
DISTILLATION
COLUMN

MTBE

Fig. 13–1 MTBE plant

The feed consists of isobutylene, fresh methanol, and recycled methanol. The isobutylene comes mixed with other C_4's (normal butylenes, iso-, and normal butane). As in Figure 13–1, the feed is charged to a fixed bed reactor and passes through the catalyst bed, indicated by the X. The solid catalyst, an acidic ion-exchange resin, sits loosely in the vessel to allow easy passage but intimate feed/catalyst contact. The combination of only moderate temperatures and the catalyst promotes the reaction between the methanol and the isobutylene. The reaction takes place at 120–200°F and 300 psi. It is slightly exothermic, and heat needs to be removed to keep the temperature below 210°F, or by-products will abound. About 90% of the isobutylene converts to MTBE in this reactor.

The effluent from this fixed bed reactor, both vapor and liquid, goes to the catalytic distillation column where the reaction continues, converting almost all the remaining isobutylene as the gaseous C_4's, and methanol rises through the catalyst. The catalyst in this vessel is loaded in bales, sometimes called "Texas teabags." As the reaction proceeds, MTBE, a lower boiling point liquid than the C_4's and methanol, drops out of the bottom of the column as a liquid. The process is run with excess methanol, so the tops include a vapor mixture of the unconverted methanol and the other C_4's.

The easiest way to separate the methanol is to trickle some water through the mixture. Since methanol has an affinity for water but the C_4's have an aversion to it, the water/methanol mixture comes out the bottom of the methanol extraction column and the C_4's out the top. The methanol and water are separated by a simple distillation with the water recycled back to the methanol extraction column, the methanol recycled back to the beginning of the process.

Other variations of the process use sulfuric acid as a catalyst and achieve a 90% conversion of i-$C_4^=$ to MTBE. Adding a second reactor in series will boost the yield to 98%.

Commercial aspects

About 95% of the MTBE is used as an octane booster/oxygenate in gasoline, as of this writing. The rest is used is used as a solvent and as feedstock in the production of methacrolein and methacrylic acid.

MTBE Properties	
Molecular weight	88.15
Melting point	-163.5°F (-108.6°C)
Boiling point	131.5°F (55.3°C)
Specific gravity	0.741
Weight per gallon	6.18 lbs./gal.

MTBE is miscible in water. If gasoline with MTBE leaks from a storage tank, the oil will lie on top of any groundwater or underground water source but the MTBE will leach out of the gasoline and dissolve in the water. MTBE is unpleasant smelling and tasting and an alleged carcinogen. It is irritating to the eyes and throat. It has a low flashpoint and is susceptible to explosion. Neat MTBE is stored in airtight stainless steel or aluminum tanks.

Chapter 13 in a nutshell ...

MTBE is made by catalytically reacting methanol and isobutylene. The outlook for MTBE is murky because the primary application, a gasoline octane enhancer/oxygenate, is under attack because of environmental reasons.

EXERCISES

1. If the value of isobutylene in refining is $1 a gallon, and MTBE in gasoline is worth $1.40 a gallon, what can a refiner afford to pay, on the margin, for methanol feed to his MTBE plant? Assume the operating costs are 10 cents per gallon of product, the plant produces 98% yield based on isobutylene and the by-products are worthless. You also need the following information.

	Sp. gr.	Lbs./gal	Molecular weight
i-C$_4$$^=$	0.690	5.76	56.10
MeOH	0.7924	6.59	32.04
MTBE	0.741	6.18	88.15

2. What are the pros and cons of using MTBE in gasoline?

CHAPTER 14

Some Other Alcohols

*"'Twas a woman that drove me to drink,
but I never had the courtesy to go back and thank her."*

W. C. Fields, 1879–1946

There are many other commercial alcohols besides methanol. This chapter treats the ones traded in the largest volumes: ethyl alcohol, isopropyl alcohol (IPA), normal butyl alcohol (NBA), 2-ethyl hexanol (2-EH), and 1,4-butanediol (BDO).

A good way to think about alcohols is to start with water, H_2O, which can be written H-OH. An alcohol is formed if the H is replaced by an organic grouping. The chemist's way of referring to any organic grouping like a chain or a ring is the symbol *R*. So, the alcohol signature is R-OH.

Often, but not always, the alcohol is named after whatever *R* is. CH_3-CH_2-OH is ethyl alcohol, $CH_2 = CH$-OH is vinyl alcohol, but C_6H_5-OH, a benzene ring with a hydroxyl group, is phenol. (*See* Figure 14–1.)

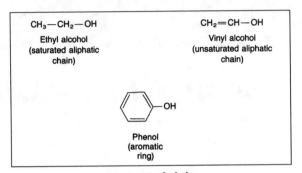

Fig. 14–1 Alcohols

By the way, when there is more than one hydroxyl group per molecule, it is still an alcohol. Ethylene glycol

$$
\begin{array}{cc}
CH_2 - & CH_2 \\
| & | \\
OH & OH
\end{array}
$$

is a polyhydric alcohol, as is glycerin and 1,4 - butanediol

$$
\begin{array}{ccc}
CH_2 - CH - CH_2 & \quad & CH_2 - CH_2 - CH_2 - CH_2 \\
| \quad | \quad | & \quad & | \qquad\qquad\qquad | \\
OH \quad OH \quad OH & \quad & OH \qquad\qquad\qquad OH
\end{array}
$$

or glucose.

$$
\begin{array}{c}
CH_2 - CH - CH - CH - CH - CHO \\
| \quad\ \ | \quad\ \ | \quad\ \ | \quad\ \ | \\
OH \quad OH \quad OH \quad OH \quad OH
\end{array}
$$

Like Caesar's Gaul, most petrochemical processes used for manufacturing alcohols today *est divida in tres partes.*

1. **Hydration.** The addition of water to an olefin:

$$CH_2 = CH_2 + H_2O \rightarrow CH_3\text{-}CH_2\text{-}OH$$
$$\text{Ethylene} \qquad\qquad \text{Ethyl alcohol}$$

2. **Oxo reaction.** Reacting an olefin with synthesis gas (CO and H_2) to produce an aldehyde (called hydroformylation) followed by hydrogenation (addition of hydrogen), producing an alcohol containing one more carbon than the original olefin.

$$\text{Olefin} + \text{Syngas} \rightarrow R\text{-CHO}$$

then,

$$R\text{-CHO} + H_2 \rightarrow R\text{-CH}_2\text{-OH}$$

3. **Ziegler reaction.** Producing an even number, straight-chain alcohol in carbon number range C_8 to C_{18}. The reaction involves "growing" chains of ethylene on an aluminum-organic compound (they grow as three branches), oxidation of the trialkyl aluminum, then water hydrolysis to "clip off" the alcohol.

$$Al(C_2H_5)_3 + CH_2 = CH_2 \rightarrow \text{Trialkyl aluminum}$$
$$\text{Triethyl aluminum} \qquad \text{Ethylene} \qquad \text{(A polymer chain)}$$

then,

$$H_2O$$
$$\text{Trialkyl aluminum} + O_2 \rightarrow 3R\text{-OH} + Al(OH)_3$$
$$\text{Alcohol} \qquad \text{Aluminum hydroxide}$$

where *R*, in this case, is a C_8 up to C_{18} alkyl group.

The best way to elaborate on these processes is to look at specific alcohols.

ETHYL ALCOHOL

The fermentation of sugar in the presence of yeast to produce ethyl alcohol in the form of wine goes back beyond historians' recorded words. The sugar came from grapes. Later, starch from grain, potatoes, or "corn squeezins" was used also. The yeast came from living matter in the form of mold or fungus. Yeast contains the enzyme "zymase." It's this enzyme that catalyzes the fermentation of sugar. Mix sugar (in grape juice) with yeast, and they will react slowly—weeks, months, maybe years, to form ethyl alcohol and carbon dioxide, as well as minor amounts of some aldehydes. Depending on preferences, some of the nonalcoholic contents can be separated by distilling.

Alcoholic beverages in the United States are made exclusively by the fermentation process not the petrochemical process. It has nothing to do with the chemistry. It's a law enacted to protect the grain growers not the consumers.

The convention for identifying the alcoholic content of beverages is "proof." So, 100 proof is 50% ethyl alcohol; 86 proof scotch is 43% ethyl alcohol, and so on. Divide by two. So pure ethyl alcohol is 200 proof.

Until World War I, fermentation accounted for all the ethyl alcohol produced in the United States. In 1919, a petrochemical route based on ethylene, sulfuric acid, and water was developed commercially and called indirect hydration. By 1935, only 10% of the ethyl alcohol was produced this way, primarily because of the expense of the ethylene at that early stage of the industry. With the rapid improvements of ethylene technology, the share quickly grew to 90% by the 1960s. At that time, an alternate route, direct hydration, was developed eliminating the use of sulfuric acid and one step in the process. Direct hydration replaced the indirect hydration process by the 1970s. Advantages were higher yields, less pollution, and lower plant maintenance due to less corrosion—all leading to better economics. Currently, almost all the domestic ethyl alcohol is produced via the direct catalytic hydration of ethylene.

The process

The chemical reaction,

$$CH_2 = CH_2 + H_2O \rightarrow CH_3\text{-}CH_2\text{-}OH$$

takes place in a single reactor, as shown in Figure 14–2. The rest of the facilities are handling and cleanup hardware.

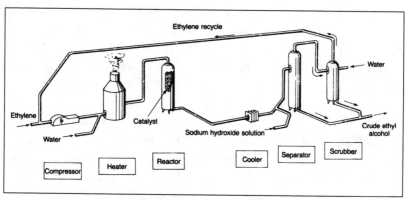

Fig. 14–2 Ethyl alcohol plant

Ethylene is compressed to 1000 psi, mixed with water, and heated to 600°F. The two reactants, both in a vapor phase, are fed down a catalyst-filled reactor. The catalyst is phosphoric acid (H_3PO_4) absorbed onto a porous inert support (usually diatomaceous earth or silica gel).

The ethylene conversion to ethyl alcohol per pass through the reactor is only 4–6%, so most of the ethylene needs to be recycled. But first the reactor effluent is cooled and caustic washed to neutralize any vaporized H_3PO_4. As the effluent cools down, the ethyl alcohol liquefies, and the ethylene can easily be separated. The ethylene recycle stream is then "scrubbed" by sloshing it through water prior to recycle.

The mixture from the bottom of the separator and the scrubber is crude ethyl alcohol. That is, it contains the ethyl alcohol, water, and all the by-products. Further distillation separates out an ethyl alcohol-water mixture (95% ethyl alcohol, 5% water) that boils at a single, constant temperature, called an azeotrope. Now that presents a special, knotty problem. Since the mixture boils at a temperature lower than ethyl alcohol, how does the ethyl alcohol get separated from the water? Not by ordinary distillation.

The answer is like fighting fire with fire—another azeotrope is formed. When benzene is added to ethyl alcohol and water a ternary azeotrope, a mixture of three compounds that boil at a single temperature, is formed. The ternary azeotrope has the composition of 68% benzene, 24% ethyl alcohol, and 6% water, and it boils at a temperature lower than the binary ethyl alcohol/water azeotrope. So, when a little benzene is added to the ethyl alcohol/water mixture and then put through a distillation column, the ternary azeotrope, in a 68-24-6 composition will come off the top, taking with it all the benzene, all the water, but just some of the ethyl alcohol. Out the bottom comes what's left, the rest of the ethyl alcohol in nearly pure form. Slick. None of this, by the way, is shown in Figure 13–2.

The ternary azeotrope is liquefied, which causes it to phase separate. That is, it separates into two layers of liquid, one of benzene plus ethylene alcohol and one of water. The benzene/ethyl alcohol is drawn off and split in another column to create pure ethyl alcohol and a benzene recycle stream.

By-products often mixed with the ethyl alcohol are diethyl ether (an anesthetic) and acetaldehyde, both of which can be easily hydrogenated to ethyl alcohol.

Material Balance

Feed:

Ethylene	640 lbs.
Water	412 lbs.
Catalyst	small

Product:

Ethylene alcohol	1000 lbs.
By-products	52 lbs.

Alternative routes

A route to ethyl alcohol based on methanol chemistry is grabbing a small but increasing share of the pie. The process has three steps. The methanol is reacted with carbon monoxide in the liquid phase in the presence of a catalyst to form acetic acid.

$$CH_3OH + CO \rightarrow CH_3COOH$$

The acid is then esterified with methanol to methyl acetate.

$$CH_3COOH + CH_3OH \rightarrow CH_3 - \overset{\overset{\displaystyle O}{\displaystyle \|}}{C} - OCH_3 + H_2O$$

The methyl acetate is then hydrogenated to ethyl alcohol (and methanol).

$$CH_3 - \overset{\overset{\displaystyle O}{\displaystyle \|}}{C} - OCH_3 + 2H_2 \rightarrow CH_3CH_2OH + CH_3OH$$

Commercial aspects

Uses. Nearly half the ethyl alcohol produced in petrochemical plants (not the stuff fermented for human consumption) is used as a chemical intermediate in the manufacture of ethyl acrylate, ethyl amines, ethyl acetate (when you pop the cap on nail polish remover, you smell ethyl acetate), ethylene chloride, glycol ethers, acetaldehyde, and acetic acid. However, you will see in the chapters on acetaldehyde and acetic acid, there are now more competitive routes than those based on ethyl alcohol.

The balance of the industrial ethyl alcohol is in demand as a solvent in personal care products (aftershave lotion, mouthwash), inks, cosmetics, detergents, household cleaners, pharmaceuticals, industrial coatings, and as a processing solvent.

Ethyl alcohol is being used extensively in the United States as an automotive gasoline supplement, known as gasohol. However, the source of the ethyl alcohol is not the petrochemical industry but the fermentation industry. Special tax incentives have been given to manufacturers of ethyl alcohol made from grains or corn, and that has made the process competitive with oil base gasoline. Those incentives are not available to petrochemical sources of ethyl alcohol, and the latter route will remain noncompetitive. Even if oil prices increase, ethyl alcohol feedstock costs are likely to increase simultaneously and in proportion, leaving the gasoline market economically unreachable by ethyl alcohol from petrochemical sources.

Properties and handling. Ethyl alcohol is a colorless, flammable liquid (good for flambé') having a characteristic odor nearly universally recognizable. It is soluble in water (and club soda) in all proportions. It's commercially available as 190 proof (the 95% ethyl alcohol-water azeotrope) and "absolute" (200 proof). It is frequently denatured to avoid the high tax associated with 190 and 200 proof grades. Methanol and/or sometimes formaldehyde are common denaturants used to prevent consumption as an alcoholic beverage.

Because of the flammability, ethyl alcohol is transported as a hazardous material.

Ethyl Alcohol Properties	
Molecular weight	46.07
Boiling point	179.2°F (78.3°C)
Freezing point	-137.9°F (-114.1°C)
Specific gravity	0.789 (lighter than water)
Weight per gallon	6.58 lbs./gal.

ISOPROPYL ALCOHOL

Indirect hydration, the traditional route, took advantage of readily available refinery grade propylene and cheap sulfuric acid in a quick two-step to isopropyl alcohol. Persistent catalysis research has now resulted in a direct route involving a small amount of an arcane catalyst, less energy intensity, high conversion rate, and an overall cheaper process.

Indirect hydration

The specifications for the feed to the indirect hydration route to IPA plant can be loose. Refinery grade propylene, even with some small amounts of ethane and ethylene can be used, because the C_2's and propane don't react. They just pass through the process. As a matter of fact, the process acts as kind of a C_3 splitter, since about 50% of the propylene gets converted to IPA in each pass through the reactor, leaving high purity propane behind.

Propylene is absorbed by concentrated sulfuric acid to form isopropyl hydrogen sulfate. That's subsequently hydrolyzed with water to IPA and dilute sulfuric acid.

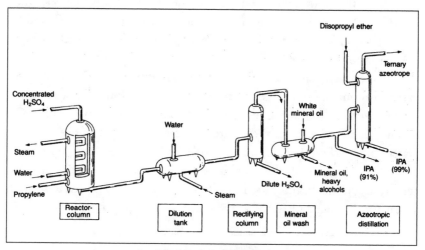

Fig. 14–3 Isopropyl alcohol plant

The propylene stream is fed into the bottom of a column (*see* Figure 14–3) packed with baffles to give intimate contact. Sulfuric acid in concentration as low as 65% is introduced at the vessel top.

As the acid and propylene slosh past each other, about 50% of the propylene reacts with the sulfuric acid to form the sulfate. The reaction is exothermic, so the contents of the tower must be continually cooled to maintain a 70–80°F temperature. This minimizes by-products, particularly propylene polymers. Any higher olefins, usually C_4 and C_5, in the propylene feed will be absorbed by the sulfuric acid, forming sulfates and bisulfates. They have to be removed in the cleanup facilities. The yield of IPA from propylene, that is the proportion of propylene that ends up as IPA, is about 70%.

The propylene from the reactor top can be recycled to the feed (though it's not shown that way in Figure 14–3.) The concentration might have to be boosted by a splitter or by the addition of some chemical grade propylene. The effluent from the reactor bottom is dumped into a lead-lined tank and diluted with water and steam, cutting the unreacted sulfuric acid to about 20%. Mixing sulfuric acid and water is exothermic, and that heat plus a little steam is sufficient to hydrolyze the isopropyl hydrogen sulfate to IPA. With a little more steam, the crude IPA flashes (vaporizes) out of the dilution tank and goes to a fractionator for concentration. The dilute H_2SO_4 stream is sent off to be cleaned up and reconstituted to higher concentration for reuse.

At the fractionator, a 91% IPA/water azeotrope distills from the top, carrying along most of the other organics. The IPA/water azeotrope is washed with mineral oil, a heavy hydrocarbon that absorbs the C_4's, C_5's, and higher alcohols. It's further treated with sodium hypochlorite to give the water-white technical grade, which is still the 91% IPA/water azeotrope.

Like ethyl alcohol, the absolute (99⁺%) grade is made by forming a ternary azeotrope. In this case, DIPE (di-isopropyl ether) is used to form the ternary with water and IPA. But the idea is exactly the same.

Material Balance

Feed:

Propylene oxide	900 lbs.
Water	385 lbs.
Sulfuric acid (85%)	1235 lbs.

Product:

Isopropyl alcohol	1000 lbs.
Sulfuric acid	1235 lbs.
By-products	285 lbs.

Direct hydration

IPA could always be made by direct hydration, but the severe operating conditions (high pressures and temperatures) and puny yields had always limited the economic enthusiasm for the process. Then catalysis research paid off with the development of a sulfonated polystyrene cationic exchange resin catalyst, a mouthful in itself. The breakthrough permitted reduced pressures and temperatures without loss of yield. The catalyst works in the vapor phase, the liquid phase, and the mixed phase.

Chemical grade propylene (90–92% concentration) and water are heated under pressure to 350–375° F. The partially liquid reactants are fed to the top of a reactor containing a packed bed of catalyst. Hydration to IPA occurs as the propylene water mixture trickles down the reactor through the catalyst.

The mixture leaving the bottom of the reactor consists of unreacted propylene and water, IPA, and diisopropyl ether (DIPE). In subsequent steps it is cooled, depressurized, and waterwashed. The unreacted propylene and by-product DIPE are flashed off and separated in a propylene recovery column. The unreacted propylene is compressed and recycled.

IPA in concentrations of 91% or 99% is recovered in the same manner described in the indirect hydration route. Approximately 5% DIPE forms as a by-product in this process and comes out the bottom of the propylene recovery column.

Material Balance	
Feed:	**lbs.**
Propylene (@100% purity)	716
Water	614
	1330
Product:	
Isopropyl alcohol	1000
By-products	46
Water	284
	1330

You may wonder why both these processes produce isopropyl alcohol instead of (normal) propyl alcohol. With the exception of ethylene, direct or indirect hydration of an aliphatic olefin always produces an alcohol with the hydroxyl group preferentially attached to the double-bonded carbon with the least number of hydrogen atoms.

Commercial aspects

Uses. In 1980, more than 50% of the IPA produced was used to make acetone (diethyl ketone). By the year 2000, the percent was down to less than 6%. The cumene plants that coproduce phenol and acetone had almost entirely replaced the IPA-to-acetone route, eliminating the need for IPA feed.

Now, IPA is used primarily as a coating and processing solvent in paints, electronics applications, synthetic resins, personal care products, and cosmetics. It is also used as a chemical intermediate for isopropyl esters, isopropyl amines, methyl isobutyl ketone, diisobutyl ketone, and hydrogen peroxide production.

At one time, IPA was used as a gasoline additive to prevent cold weather stalling, but it has been largely displaced by DIPE.

And of course, IPA is used as rubbing alcohol, because of its innocuous nontoxic odor, its low boiling (vaporization) temperature, and moderate heat of vaporization. It "dries" rapidly but won't give you frostbite like liquid butane might.

Properties and handling. IPA is a colorless, flammable liquid with that characteristic, rubbing alcohol odor. It's soluble in water in all proportions, as well as most organic solvents. It is commercially available in technical grade (91%), chemical (98%), and absolute (99+%). Shipments by rail, truck, drum, etc., are routine, except that the flammability requires hazardous materials warnings.

Isopropyl Alcohol Properties	
Molecular weight	60.10
Boiling point	180.5°F (82.5°C)
Freezing point	-129.1°F (-89.5°C)
Specific gravity	0.785 (lighter than water)
Weight per gallon	6.55 lbs./gal.

NORMAL BUTYL ALCOHOL AND 2-ETHYL HEXANOL

There's another convention that has been useful in describing alcohols, and that has to do with the positioning of the -OH or hydroxyl group. There are primary, secondary, and tertiary alcohols, depending on whether the hydroxyl group is attached to the primary, secondary, or tertiary carbon atom. In the case of the C_4 alcohols, the hydroxyl group can be connected to either:

- a primary carbon atom, one which is attached to only one other carbon atom
- a secondary carbon atom, one which is attached to two other carbon atoms
- a tertiary carbon atom, one which is attached to three other carbon atoms

In other words, for the C_4 alcohols you can have:

$$CH_3 - CH_2 - CH_2 - CH_2 - OH$$

Normal butyl alcohol (NBA)
($1°$ alcohol, or $R - CH_2 - OH$)

$$CH_3 - CH - CH_2 - CH_3$$
$$|$$
$$OH$$

Secondary butyl alcohol (SBA)
($2°$ alcohol, or $R - CH - OH$)
$$|$$
$$R$$

$$CH_3$$
$$|$$
$$CH_3 - C - OH$$
$$|$$
$$CH_3$$

Tertiary butyl alcohol (TBA)
($3°$ alcohol, or $\quad R \quad$)
$$|$$
$$R - C - OH$$
$$|$$
$$R$$

Normal butyl alcohol (NBA) was first recovered in the 1920s as a by-product of acetone manufacture via cornstarch fermentation. That route is almost extinct now. A small percent is still made from acetaldehyde. The primary source of NBA, however, is the Oxo process.

The Oxo process is used in a number of applications for extending the length of an olefin chain by one carbon. The reaction is between an olefin and synthesis gas (carbon monoxide and hydrogen, covered just in time in the last chapter) in the presence of a cobalt or, more recently, a rhodium catalyst. It produces a mixture of aldehydes (the -CHO signature group), which readily undergo hydrogenation to alcohols. One important feature of the process is that it produces only primary alcohols. Most other petrochemical processes yielding alcohols produce secondary alcohols. (*See* Figure 14–4.)

$$CH_3 - CH = CH_2 + H_2 + CO \longrightarrow \text{Normal butyraldehyde} + \text{Isobutyraldehyde}$$

$$\downarrow \text{add } H_2$$

$$\text{Normal butyl alcohol} + \text{Isobutyl alcohol}$$

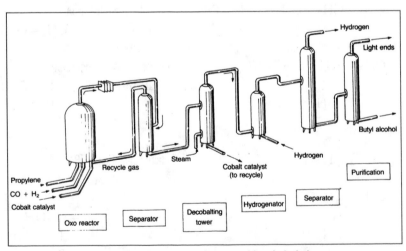

Fig. 14–4 Oxo process for normal butyl alcohol

Both of the iso- and normal aldehydes and alcohols are formed in this process, but the demand for NBA greatly exceeds the isomer. By using mild operating conditions in the plant, the ratio of normal to iso- increases, though the total yield goes down. Recycling the isobutyraldehyde to the feed "fools" the reaction and also keeps down the iso- yield. Most processes yield an overall ratio of normal to iso- of about 4 to 1.

2-Ethyl hexanol, sometimes called 2-ethyl hexyl alcohol, 2-ethyl hex, or more simply 2-EH, is one of the oldest high molecular weight aliphatic alcohols. What does it have in common with NBA? Both are made from propylene via the Oxo process, and both have the same aldehyde intermediate—normal butyraldehyde.

In the case of 2-EH production, the aldehyde dimerizes or reacts with itself. (*Dimerize*, remember, has the same Latin root -*meros*, as isomer, monomer, and polymer, and means part. A dimer is a chemical union of two molecules of the same compound.) The resulting C_8 dimer is also an aldehyde that can be hydrogenated to give 2-EH.

Normal butyraldehyde ⟶ "Dimer" aldehyde

$$\downarrow \text{ add } H_2$$

$$CH_3 - CH_2 - CH_2 - CH_2 - CH - CH_2 - OH$$
$$|$$
$$CH_2 - CH_3$$

The name 2-EH becomes apparent from the layout of the molecule. The straight C_6 chain with the -OH on the end gives the hexanol; the group, -CH_2-CH_3, gives the ethyl; and the two comes from ethyl group being connected to the second carbon from the functional -OH group.

The process for NBA

To make, butyraldehyde, the precursor for NBA, the so-called Oxo process is used, reacting chemical grade propylene with hydrogen and carbon monoxide at 250–300°F and 3500–4000 psi. Under those conditions, both feeds are liquids. The catalyst is an oil-soluble cobalt carbonyl complex dissolved in the propylene. If rhodium-based catalysts or complexes based on rhodium carbonyls and triphenyl phosphine are used, the reaction conditions are less severe (200°F and pressure of 100–350 psi) and the yields of n-butyraldehyde are slightly higher.

The reactor effluent contains unreacted gases, catalyst, and the aldehydes. The pressure is let down in a separator and the unreacted gases are recycled.

The hydrogenation step takes place in the conventional way in vessel packed with catalyst where the aldehydes and hydrogen are admixed at 200–300°F and 600–1200 psi. The catalyst is usually nickel or copper chromite on an inert carrier such as kieselguhr, silica gel, or alumina. The crude butyl alcohols are finally separated and purified by distillation.

The process for 2-EH

The same front-end Oxo process is used to make the butyraldehyde feed for 2-EH manufacture, but then the butyraldehyde is dimerized in a reaction called aldol condensation. Some plants even combine the Oxo process and the Aldol process and then refer to it as the Aldox process.

The dimerization (Aldol condensation) takes place at temperatures of 175–250°F in the presence of a dilute solution of sodium hydroxide. After the reaction, the mixture is passed to a separator tank where the dimer is separated then sent to a reactor to be hydrogenated over a nickel catalyst. Reaction conditions have temperatures of 300°F and 2500 psi. Distillation of the effluent gives purified 2-EH in 95% yield.

It used to be that the iso-butyraldehyde had to be split out before the dimerization reaction, but catalyst improvements have permitted cogeneration of both C_4 alcohols and 2-EH.

Commercial aspects

Uses. The motivation for first recovering NBA in the 1920s was its use as a lacquer solvent. That application is even stronger today. The NBA vapors from lacquer drying are nontoxic and virtually nonflammable. Other fast growing uses for NBA are plasticizers and chemical intermediates, mostly for esters and ethers used in water-based coatings and adhesives systems.

The most important use for 2-EH is in the manufacture of di-2-ethyl hexyl phthalate (also known as dioctyl phthalate)—neither name would you want to try late on a Saturday night—which is used as a plasticizer to make polyvinyl chloride flexible. About 65% of the 2-EH goes to the manufacture of plasticizer. The growth of 2-EH acrylate is becoming an important application for 2-EH. This acrylate is finding use as a construction adhesive and in surface coatings. Other uses include industrial solvents, dispersing and wetting agents, and chemical intermediates.

Properties and handling. NBA and 2-EH are nonvolatile, colorless, nontoxic liquids, with relatively high boiling points. NBA is only slightly soluble in water and 2-EH is insoluble. A little rule of thumb here on solubility: "Like dissolves like." Methanol is very much like water, because the hydroxyl group in both (CH_3-OH and H-OH) is a significant part of the molecule. The same is true of ethyl alcohol and IPA. But as you get up to NBA, and especially 2-EH, the hydroxyl is minor and the carbon chain significant. The analogy between compounds and other solvents, including organics, usually holds true—"Like dissolves like."

Both NBA and 2-EH are available in technical grade (98–99%) and are transported in normal equipment. No hazardous material label is required.

Normal Butyl Alcohol Properties

Molecular weight	74.12
Freezing point	-129.1°F (-89.5°C)
Boiling point	243.1°F (184.3°C)
Specific gravity	0.811 (lighter than water)
Weight per gallon	6.75 lbs./gal.

2-Ethyl Hexanol Properties

Molecular weight	130.23
Freezing point	-94.0°F (-70.0°C)
Boiling point	363.7°F (184.3°C)
Specific gravity	0.834 (lighter than water)
Weight per gallon	6.94 lbs./gal.

SECONDARY AND TERTIARY BUTYL ALCOHOLS

Unfortunately, secondary and tertiary butyl alcohols (SBA and TBA) cannot be made by the Oxo process. Instead they are produced either by indirect or direct hydration of the corresponding olefin. Normal butylene gives SBA and isobutylene gives TBA. The processes are similar to the corresponding routes to IPA.

SBA is a colorless, high boiling point (212°F) liquid with a pleasant odor. TBA, on the other hand, is a white solid (melting point is 78°F) with a camphor-like odor. Both alcohols are traded as technical grade (99% purity) and need a hazardous (corrosive) materials label.

SBA is primarily used as feedstock for methyl ethyl ketone. Other uses include hydraulic fluids, industrial cleaning compounds, paint remover, and an extracting agent for oils, perfumes, and dyes. TBA is used mostly as feedstock to make methyl methacrylate and glycol ethers (by reaction with ethylene or propylene oxides.) TBA is also a coproduct with PO, as covered in Chapter 11.

1,4-BUTANEDIOL

The most recent entrant to the club of commodity chemicals is 1,4-butanediol (BDO), a petrochemical used in some of the more specialized applications such as chemical intermediates for the production of tetrahydro-furane and gama-butyrolactone, polybutylene terephthalate, and the more familiar polyurethanes. Traditionally, the Reppe process was the primary route to BDO, based on acetylene and formaldehyde feeds. More recently, the share of BDO from butane and propylene oxide based production has grown rapidly.

In the Reppe process, formaldehyde and acetylene are reacted in the presence of a copper acetylide catalyst to give 2-butyne-1,4-diol. That compound is then hydrogenated to give BDO.

$$2CH_2O + HC\equiv CH \longrightarrow HO-CH_2-C\equiv C-CH_2-OH \quad \text{(2-butyne-1,4-diol)}$$

Plus H_2

$$HOCH_2-CH_2-CH_2-CH_2OH \quad \text{(1,4-butanediol)}$$

Acetylene is a tricky chemical feedstock. It is extremely reactive (and explosive) and impractical to transport. Generally, the industrial processes that use acetylene are close to the acetylene-generating source. Despite all the drawbacks, the Reppe process was for a half century the preferred process for BDO, but now the growth in BDO is being taken by the propylene oxide (PO) and butane feedstock routes.

In the PO route, PO is isomerized to allyl alcohol in the presence of a lithium phosphate catalyst.

$$\overset{\displaystyle O}{\overset{/\ \ \backslash}{CH_3-CH-CH_2}} \rightarrow CH_2{=}CH-CH_2OH$$

The allyl alcohol is purified by distillation and is hydroformylated with syngas in the presence of a rhodium catalyst to give hydroxybutyraldehyde.

$$CH_2{=}CH-CH_2OH + CO + H_2 \rightarrow HOCH_2-CH_2-CH_2-CHO$$

The hydroxybutyraldehyde is subsequently hydrogenated with a Raney nickel catalyst to give BDO.

$$HOCH_2-CH_2-CH_2-CHO + H_2 \rightarrow HOCH_2 - CH_2 - CH_2 - CH_2OH$$

In the butane route, a chemically complicated three-step process is needed to get from the feed to BDO. The two feeds, oxygen (air is used) and butane, are fed to a fluid bed reactor admixed with a catalyst. In a fluid bed reactor, the feeds and catalyst move continuously and, in this case, at a uniform temperature that allows optimum conditions for the catalyst to do its work. Butane and oxygen react to form maleic anhydride (MA), a cyclic compound. The fixed bed reactor effluent gases are taken off overhead, cooled, and filtered to remove entrained catalyst particles. The gases are then moved to a scrubber where the MA is then hydrolyzed to maleic acid, a straight chain molecule. This is a relatively simple reaction because of the reactivity of the maleic anhydride ring. Scrubber overhead gases are incinerated for safe disposal. The maleic acid is then catalytically hydrogenated in a fixed bed reactor to give BDO in yields of 94% or more.

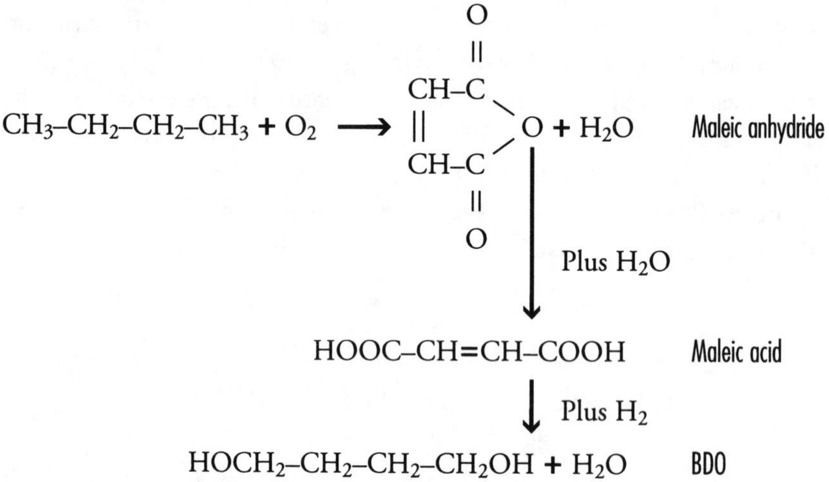

$$CH_3-CH_2-CH_2-CH_3 + O_2 \longrightarrow$$

Maleic anhydride

Plus H_2O

$$HOOC-CH=CH-COOH$$ Maleic acid

Plus H_2

$$HOCH_2-CH_2-CH_2-CH_2OH + H_2O$$ BDO

Properties and handling. 1,4-butanediol is a colorless, oily liquid with a feint odor. It is miscible in water, ethanol, and acetone. BDO is not a skin irritant but it is toxic if you drink it. It is available as technical grade. There are no bulk shipping regulations.

1,4-Butanediol Properties	
Molecular weight	90.0
Freezing point	60.8°F (16°C)
Boiling point	446.0°F (230°C)
Specific gravity	1.02
Weight per gallon	8.6 lbs./gal.

Commercial aspects

The growing market for birth control devices is driving the demand for tetrahydrofurane while mothers' milk substitutes create the gamma-butyrolactone demand. Not really. Just checking to see if you're reading all this. Polybutylene terephthalate is an engineering thermoplastic being used in automobile and electrical components. Other minor uses include solvents, humectant, plasticizer, and pharmaceuticals.

Chapter 14 in a nutshell ...

Alcohols have the characteristic -OH signature group. There are a variety of ways to get that signature affixed to various hydrocarbons: (a) synthesis gas is the source of oxygen and hydrogen as well as some of the other parts in methanol, normal butyl alcohol, and 2-ethyl hexanol; (b) ethylene can be directly hydrated to ethanol (ethyl alcohol); (c) isopropyl alcohol is now made by either direct or indirect hydration of propylene; and (d) old technology for 1,4-butanediol based on acetylene is giving way to routes based on propylene oxide or butane/air, both using exotic catalysts.

EXERCISES

1. Why can't pure ethyl alcohol be obtained by simple fractionation of ethyl alcohol and water? What is a "constant boiling mixture?" Give four examples from this chapter and one from a much earlier chapter. (Hint: use the Index.)

2. Why should normal hexyl alcohol be less soluble in water than isopropyl alcohol?

3. Approximately 2.6 gallons of ethyl alcohol (95% ethyl alcohol and 5% water) are produced per bushel of corn. Fermentation costs about 50 cents per gallon of ethyl alcohol, and corn is selling for about $2.50 per bushel. If the operating costs of an ethyl alcohol plant are 30 cents per gallon, what is the maximum price that could be paid for ethylene to remain competitive with "natural" ethyl alcohol?

4. Which alcohols are made commercially by direct hydration and which ones can only be made by indirect hydration or more complicated processes? Which ones can go either way?

CHAPTER 15 ⬡

The Higher Alcohols

"For the proverbe seith that 'manye smale maken a greet.'"

The Canterbury Tales
Geoffrey Chaucer, 1343–1400

Higher than what? Well, the higher alcohols start with C_6 and go to C_{18} plus. They are predominantly straight chains and they are used to make plasticizers and detergents. Consequently they are often referred to as *plasticizer* alcohols (the C_6–C_{12} alcohols) and *detergent* alcohols (C_{12}–C_{18} alcohols). The detergent alcohols are also sometimes called the fatty alcohols, but that has to do with their origin, not their destination.

These higher alcohols can come from any of three processes. All three are in commercial use today.

- Nature's way—hydrogenation of fatty acid esters derived from natural fats and oils, mainly coconut oil, palm kernel oil, and tallow
- Ziegler process—ethylene oligomerization
- Oxo process—the conversion of internal or alpha olefins, alpha meaning that the double bond is located between the number one and two carbon atoms, i.e., at the end of the chain

Nature's way

The origins of the detergent alcohol business are the connections between natural fats and oils and the surfactants that end up in shampoos and other detergents. The major ingredient in fats and oils such as coconut oil and palm kernel oil is the naturally occurring ester, fatty triglyceride in Figure 15–1. This awkward-looking molecule is glycerol, esterified with (converted to esters by) three normal, straight-chain fatty acids. Fatty acids are so-called, again, because they are derived originally from animal fats or vegetable fat or oil. They contain a chain of alkyl groups and a carboxyl group, -COOH or

$$-\overset{O}{\overset{\|}{C}}-OH$$

In Figure 15–1, the R_1, R_2, and R_3 are carbon chain lengths ranging from C_4–C_{22} or more, though in these fats, C_{11}–C_{17} predominate. Batches of triglycerides from various sources have their own random distribution of fatty acid chain lengths, but from the same fats and oils the distributions are fairly consistent.

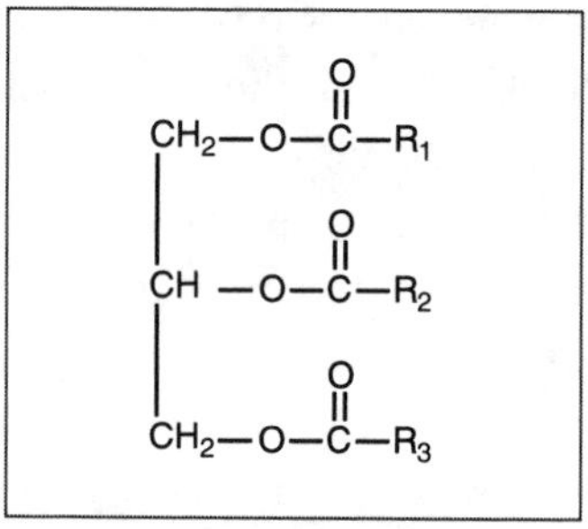

Fig. 15–1 Triglyceride

To get to the end product, the higher alcohol, the triglycerides go through a two-step process. First, the branches are split from the triglyceride in an ester interchange by reacting with methanol. Other alcohols are sometimes used but the methanol turns up again later in the process and is relatively easy to separate from the other reactants.

An alkaline catalyst, sodium methoxide, is created in the same reactor as the ester interchange by first adding sodium metal to the methanol before the triglyceride is introduced. The reaction is performed batch by batch in a carbon-steel vessel at atmospheric pressure. The reaction, shown in the top part of Figure 15–2, results in glycerol and three fatty methyl esters.

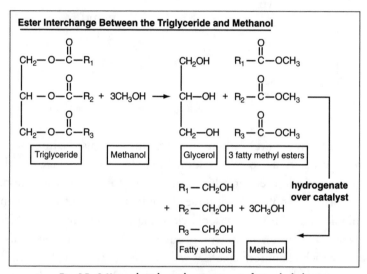

Fig. 15–2 Natural triglycerides to esters to fatty alcohols

To continue the process, the fatty methyl esters are phase-separated from the glycerin (or glycerol—same thing, just to keep you on your toes), washed with water to remove any trace amounts of methanol and glycerin and dried. In a second reaction, the methyl esters are hydrogenated to get the fatty alcohols (in the southeast corner of Figure 15–2) . The catalyst is usually a mixture of cupric chromite and cupric oxide in the form of a finely divided powder. Conversion of the triglycerides is about 95%.

The methanol and catalysts are recovered for reuse. The mixed fatty alcohols are then fractionated to give the R_1, R_2, and R_3 cuts of alcohols. Keep in mind there will be three Rs on the triglyceride, but not necessarily the same three on all the triglycerides. For example, when coconut oil is the raw material, the following distribution of straight-chain primary alcohols typically occurs (see Table 15–1):

Alcohol Chain Length Weight Distribution	
Alcohol chain length	Percent, by weight
C_8	5–9
C_{10}	6–10
C_{12}	44–52
C_{14}	13–19
C_{16}	8–11
C_{18}	6–14
$C_{20}+$	~0.4

Table 15–1

Wonder why no odd carbon numbers turn up? Just Nature's way. And Nature may be trying to tell us something else because there are also no branch chain alcohols either, which calls for the following explanation.

The primary application of these alcohols is the manufacture of anionic or nonionic surfactants for personal cleansing products, most of which end up in your wastewater treatment plants and rivers. Microorganisms don't chew up branch-chain surfactants as well as they do the straight ones. It used to be, for example that the surfactant based on the sodium salt of dodecyl benzene sulfonate, a 12-carbon branch chained anionic surfactant, was found to be slowing down water treatment processes. Dodecyl alcohol as a raw material for these surfactants has been largely replaced by laurel alcohol, a 12-carbon straight-chain, linear alcohol. If you look at the bottle next time you shampoo your hair and rinse, you'll see sulfonates based on laurel alcohol listed, but none based on dodecyl.

The thought may have occurred to you that it might be easier to directly hydrogenate the triglyceride to give the fatty alcohols in one step. While that will work, the high temperatures needed for the hydrogenation cause most of the by-product glycerine to decompose. That's a waste because glycerine is worth alot more thatn the decomposed by-products.

Actually, the pioneers of large-scale conversion of triglycerides recognized this and used direct hydrolysis (instead of hydrogenation) of the triglycerides to give glycerine and fatty acids; then they hydrogenated the acids to the fatty alcohols. To their chagrin, along came this newer route using methanol to get a methyl ester exchange with the triglyceride, then hydrogenolysis to split off the alcohols and recover the methanol. That turned out to be, as they say in the information technology world, "the killer app" because of its lower costs. It required lower reaction temperatures and pressures, had less corrosion, less expensive equipment, and crushingly better economics.

Ziegler process

This route is named after our hero, the brilliant German chemist, Karl Ziegler. He found that triethyl aluminum could, under the right conditions, be used as a kind of seed for growing hydrocarbon chains. These chains can be manipulated to yield either straight-chain primary olefins or, in the application for this chapter, linear, straight-chain primary alcohols. (The olefins, referred to as alpha olefins because the double bond occurs in next to the last carbon, are covered in Chapter 21, Alpha Olefins.)

Figure 15–3 lays out the four-step process, starting with germinating the seed from which everything sprouts. Triethyl aluminum is created from aluminum, hydrogen, and ethylene in step one, which itself has several parts. Powdered aluminum in a toluene slurry is first converted to diethyl aluminum hydride, $HAl(C_2H_5)_2$, at 212–300°F and 1500 psi. This product is then fed to a tubular reactor with ethylene at 212°F and 300 psi to produce triethyl aluminum. Yields are about 90%.

Having sown the seed, germination takes place in step two. The triethyl aluminum is reacted with ethylene to grow oligomers, repeating sets of ethyl groups as the ethylene molecules add themselves on. The reaction is very exothermic, and heat must be constantly removed from the reactor to maintain a 200–212°F temperature. Because the source of the propagation is ethylene, only even-numbered chains and no side branches result. In Figure 15–3, the R represents the oligomer and the 1, 2, and 3 reflect different length sprouts.

Picking the fruit requires two steps. In step three, the triethyl aluminum product is oxidized to an aluminum trioxide alkane (trialkoxide) by adding dry air at about 100°F above atmospheric pressure. This is done with the triethyl aluminum product mixed in a solvent to avoid local overheating and decrease the viscosity of the solution. The solvents and by-products are removed overhead by distillation.

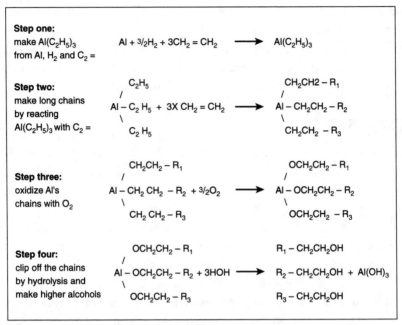

Step one:
make Al(C₂H₅)₃
from Al, H₂ and C₂ =

$$Al + 3/2H_2 + 3CH_2 = CH_2 \longrightarrow Al(C_2H_5)_3$$

Step two:
make long chains
by reacting
Al(C₂H₅)₃ with C₂ =

$$\begin{matrix} C_2H_5 \\ / \\ Al - C_2H_5 \\ \backslash \\ C_2H_5 \end{matrix} + 3X\ CH_2 = CH_2 \longrightarrow \begin{matrix} CH_2CH2 - R_1 \\ / \\ Al - CH_2CH_2 - R_2 \\ \backslash \\ CH_2CH_2 - R_3 \end{matrix}$$

Step three:
oxidize Al's
chains with O₂

$$\begin{matrix} CH_2CH_2 - R_1 \\ / \\ Al - CH_2\ CH_2 - R_2 \\ \backslash \\ CH_2\ CH_2 - R_3 \end{matrix} + 3/2O_2 \longrightarrow \begin{matrix} OCH_2CH_2 - R_1 \\ / \\ Al - OCH_2CH_2 - R_2 \\ \backslash \\ OCH_2CH_2 - R_3 \end{matrix}$$

Step four:
clip off the chains
by hydrolysis and
make higher alcohols

$$\begin{matrix} OCH_2CH_2 - R_1 \\ / \\ Al - OCH_2CH_2 - R_2 \\ \backslash \\ OCH_2CH_2 - R_3 \end{matrix} + 3HOH \longrightarrow \begin{matrix} R_1 - CH_2CH_2OH \\ R_2 - CH_2CH_2OH + Al(OH)_3 \\ R_3 - CH_2CH_2OH \end{matrix}$$

Fig. 15–3 Ziegler process for higher alcohols

In step four the bottoms are mixed with water and a trace of sulfuric acid. The chains separate from the aluminum in favor of a hydrogen atom, creating the higher alcohols and aluminum hydroxide, $Al(OH)_3$, which drops out of solution as a precipitate. The aluminum product can be readily dehydrated to alumina, Al_2O_3, and sold. Some processes use sulfuric acid, H_2SO_4, to do the hydrolyzing instead of water. That results in by-product aluminum sulfate, $Al(SO_4)_3$, a marketable but less valuable product than alumina.

The higher alcohols are finally separated into plasticizer and detergent grades and purified by distillation.

The length and the distribution of chain lengths are functions of the temperature, pressure, residence time, catalyst characteristics, and the proportion of ethylene present in the reaction. A measure of this is the "mole ratio" of ethylene, which measures the weight of ethylene compared to the weight of triethyl aluminum in scales related to their atomic weights. As an example, Table 15–2 shows how the distribution of chain lengths can vary, using different mole ratios of ethylene to triethyl aluminum.

% weight at various mole ratios of ethylene to triethyl aluminium			
Chain size	3:1	3.5:1	4:1
C_2	1.6	0.8	0.5
C_4	8.1	5.1	3.2
C_6	17.3	12.8	9.2
C_8	22.4	19.4	15.9
C_{10}	20.7	20.8	19.5
C_{12}	14.7	17.3	18.5
C_{14}	8.5	11.7	14.3
C_{16}	4.1	6.6	9.3
C_{19}	1.7	3.2	5.2
C_{20+}	0.9	2.3	4.4

Table 15–2 Typical chain lengths

Flexing the operating conditions of the versatile Ziegler process allows the distribution of the plasticizer alcohols (C_6–C_{10}) versus the detergent alcohols (C_{12}–C_{18}) to vary from as low as 15% to as high as 85%.

The Oxo process

The Oxo process, a.k.a. the hydroformylation reaction, is the same as that described in Chapter 14, Some Other Alcohols, in the section on NBA. This process can also be applied to higher alcohols by just substituting feeds of longer chain length.

The process requires two conversion steps. In the first, an olefin plus synthesis gas (carbon monoxide and hydrogen) are reacted over a cobalt or rhodium catalyst to produce two aldehydes, with one being an isomer of the other.

$$R - CH=CH_2 + CO + H_2 \rightarrow R - CH_2CH_2 - CHO + R - CH - CHO$$
$$|$$
$$CH_3$$

where R is a paraffin of any length. The two isomers are then hydrogenated to give a normal and a branched alcohol.

$$R - CH_2CH_2 - CHO \qquad\qquad R - CH_2CH_2CH_2OH$$
$$+ \qquad\qquad + 2H_2 \;\rightarrow\qquad +$$
$$R - CH - CHO \qquad\qquad R - CH - CH_2OH$$
$$\mid \qquad\qquad\qquad\qquad\quad \mid$$
$$CH_3 \qquad\qquad\qquad\qquad CH_3$$

The Oxo alcohols contain one more carbon atom than the original olefin used. They differ in two respects from the natural alcohols. First the Oxo alcohols contain either even or odd numbers of carbon atoms, the natural alcohols only even. Second, the Oxo alcohols all have some branched molecules—originally about 40% but with recent process improvements, as low as 5%. (*See* Figure 15–4.)

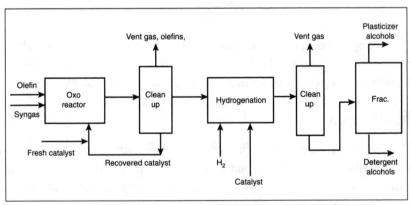

Fig. 15–4 Oxo process

The Oxo process uses synthesis gas with a H_2 to CO ratios of 1:1 to 2:1. Hydroformylation in the Oxo reactor takes place at moderate temperatures, 212–400°F, but very high pressures, 3000–5000 psi. The effluent from the reactor is cleaned up to remove light gas by-products to recover the catalyst for recycling and to recycle any unreacted olefin. At this point, the balance between the aldehyde isomers can be adjusted by fractionating out the iso-aldehyde. The rest is fed to the hydrogenation section where the alcohols are formed at 100–400°F and 3–250 psi. The resulting alcohols are fractionated to separate the plasticizer and detergent alcohols.

This process seems much simpler than the Ziegler process, and you may wonder why it has not crowded Ziegler out. The problem is the olefin feed. Where do you get a ready supply of olefins the right size to feed to the process? The answer is you have to make them, and therein lies the rub. Normal paraffins from petroleum waxes or other chemical processes provide the feedstock to a two-step process, chlorination and dehydrochlorination, which produces an olefin corresponding to the paraffin.

Another source, catalytic oligomerization, which produces the alpha olefins using the Ziegler process mentioned above, has its own treatment in Chapter 21. But then the Oxo process really only replaces steps three and four in Figure 15–3. Besides, Oxo higher alcohols still have branches.

Shell Chemical has a process that does both the Oxo reaction and hydroformylation in one step in the same reactor. They use a special catalyst, thought to be cobalt modified with a trialkyl or triaryl phosphine ligand— but they are holding this one pretty close to the vest. Overall yields are 70–80%, with straight-chain alcohols representing greater than 80%. Major by-products are paraffins that are recovered and used to make olefins and then recycled back as feed. This process can also use internal olefins (with the double-bond somewhere besides the alpha position) and yield similar normal:iso alcohol ratios.

Commercial aspects

The plasticizer-range alcohols are largely used as feedstock for production of high molecular weight diesters of phthalic, adipic, azelaic, and sulfuric acids. All these are used primarily in plasticizers for polyvinyl chloride (PVC) and other plastics. The plastics industry also uses them as additives for heat stabilization, to control the viscosity of PVC plastisols, ultraviolet absorbers, flame retardants, and antioxidants. They are also found in synthetic lubricants, agricultural chemicals, and defoamers.

A small amount, maybe 5%, of the detergent-range alcohols is consumed as such. The C_{16} and C_{18} alcohols are used extensively in the cosmetics and pharmaceutical industries as emollient additives (the heaviest of the higher alcohols are actually wax-like), intermediates for perfume and flavor components, and as a basis for creams, ointments, and suppositories. But the big markets are feedstocks for surfactant derivatives and as mechanisms to introduce long-chain hydrocarbons into chemical compounds.

Major surfactant derivatives include

- **Alcohol sulfates** $(R - OSO_3^-Na^+)$ – alcohol reacted with sulfur trioxide then neutralized with sodium hydroxide. Applications include shampoo, bar soaps, and other personal care products; laundry and dishwashing soap; textiles; and additives to emulsion polymerization.

- **Alcohol ethoxylates** $(RO - (CH_2CH_2O)_nH)$ – alcohol reacted with ethylene oxide in the presence of a base catalyst. Applications include home laundry powders and liquids, industrial cleaners, and emulsifying agent in textiles, leather, paints, paper, plastics, and pharmaceuticals.

- **Alcohol ether sulfates** $(RO(CH_2CH_2O)SO_3^-Na^+)$ – alcohol ethoxylate reacted with sulfur trioxide and then neutralized in NaOH. Applications include dishwashing liquids and shampoos, home laundry powders, and personal care products.

Other minor uses of detergent alcohols include lacquer solvent, synthetic lubricants (which need lots of different kinds of slippery long-chain molecules), antifoaming agents, herbicides, lube oil additives, and stabilizers for fire extinguisher foams.

Properties and handling. The properties of the normal, i.e., straight-chain higher alcohols vary according to the length of the molecules. For example, the higher the carbon count (chain length or molecular weight) of the alcohol, the higher the melting point and boiling point. (*See* Table 15–3.) Water solubility decreases with increasing molecular weight, and the oil solubility increases—the longer the chain, the less impact the -OH has, the more it looks like oil. Below C_{12}, the normal alcohols are colorless, oily liquids with a fruity odor. Above C_{12}, they shift to soft platelets to white waxy solids. Finally, the chemical reactivity decreases with increasing molecular weight.

	Formula	Mole. Weight	Melting Point °F	Boiling Point °F	Sp. Gr.	Weight Per Gal. in lbs.
Hexyl alc. or						
1-Hexanol	$CH_3(CH_2)_4CH_2OH$	102	-19.1	157.2	0.819	6.8
Octyl alc. or						
1-Octanol	$CH_3(CH_2)_6CH_2OH$	130	3.2	195.0	0.826	6.9
Decyl alc. or						
1-Decanol	$CH_3(CH_2)_8CH_2OH$	158	42.8	451.2	0.829	9.9
Lauryl alc. or						
1-Dodecanol	$CH_3(CH_2)_{10}CH_2OH$	186	75.2	498.2	0.836	7.0
Tetradecyl or						
1-Tetradecanol	$CH_3(CH_2)_{12}CH_2OH$	214	100.4	507.4	0.836	7.0
Cetyl alc. or						
1-Hexadeconol	$CH_3(CH_2)_{14}CH_2OH$	242	120.7	651.2	0.818	6.8
Stearyl alc. or						
1-Octadeconol	$CH_3(CH_2)_{16}CH_2OH$	270	138.2	410.0*	0.812	6.8

* @ 15 mm

Table 15–3 Straight-chain plasticizer and detergent alcohol properties

Only the lower molecular weight alcohols require a hazardous material shipping label because of the volatility. They have low flash points and are easily ignited. However, they are all combustible but nontoxic.

The alcohols are available in drums and via tank trucks and rail tank cars. The bulk shipments of the heavier higher alcohols need to be in equipment with heating coils since, as you can see in Table 15–3, the melting points are above ambient temperature conditions.

The alcohols are hygroscopic, i.e., they will absorb water, even from the atmosphere, and so they have to be protected from all sources of moisture, generally with a blanket of inert gas such as N_2.

Other factoids for each alcohol include:

- Hexyl alcohol is a colorless liquid offered in technical (90–99%) and purified (99.8%) grades
- Octyl alcohol is a colorless liquid with an aromatic odor offered in technical, chemical pure, and purified grades
- Decyl alcohol is a colorless liquid with a sweet odor available in technical and high purity grades
- Lauryl alcohol is a colorless liquid with a sweet odor available in technical and high purity grades
- Tetradecyl alcohol is a white solid available in technical grade
- Cetyl alcohol comes in white waxy flakes or solid with a faint odor and is available in technical, cosmetic, and NF grades
- Stearyl alcohol comes in white flakes and is available in commercial, technical, and USP grades

Chapter 15 in a nutshell ...

The higher alcohols cover the range from C_4–C_{22} and sometimes higher. They come from either natural sources or one of two synthetic processes, the Oxo process and the Ziegler process. Both synthetic routes involve oligomerization of ethylene, either directly (Ziegler) or indirectly (Oxo). The two major uses of the higher alcohols are plasticizers (the lower end of the spectrum) and detergents (the higher end).

EXERCISES

1. In 25 words or less, describe the differences between the Oxo process and the Ziegler process for making higher alcohols.

2. What can make the two processes very similar to each other?

CHAPTER 16 ⬡

Formaldehyde and Acetaldehyde

"A precedent embalms a principle."

Benjamin Disraeli, 1804–1881

A good warm-up to the discussion of aldehydes is a repeat and an elaboration of Chapter 1's discussion of oxidation. The chemical route to many petrochemicals is oxidation—the reaction of an atom or a molecule with oxygen. The oxygen can come from air or from another compound that readily gives up its oxygen, such as hydrogen peroxide, H_2O_2.

If oxidation is taken to its extreme, that is, complete oxidation of an organic compound, you end up with only oxygen attached to each carbon atom (CO_2) and with water H_2O. Burning (combustion) is an example of complete oxidation. Your body functions also are a good example. For instance, if vodka (ethyl alcohol) is ingested, the ultimate result is a chemical imbalance in your system:

$$C_2H_5OH + 3O_2 \rightarrow 2CO_2 + 3H_2O$$

The more vodka, then later on the more CO_2 and the less oxygen in the body. That's why you can partially relieve a hangover by breathing from an oxygen mask. It restores the normal oxygen balance to your body (and head).

In petrochemicals, partial oxidation, rather than complete, is more desirable, and gives rise to several major classes of compounds.

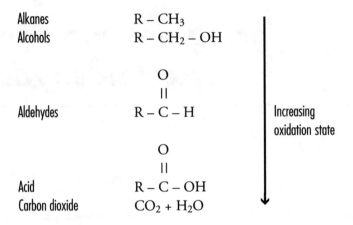

Alkanes	$R - CH_3$	
Alcohols	$R - CH_2 - OH$	
Aldehydes	$R - \overset{\overset{O}{\|\|}}{C} - H$	Increasing oxidation state
Acid	$R - \overset{\overset{O}{\|\|}}{C} - OH$	
Carbon dioxide	$CO_2 + H_2O$	

The aldehydes, specifically formaldehyde and acetaldehyde, are midway in this spectrum.

The aldehyde signature, $-\overset{\overset{O}{\|\|}}{C}-H$ (written also as -CHO, but never -COH), is always located at the end of the carbon chain. Common names for aldehydes are derived from the corresponding acid to which they are converted by further oxidation. The suffix -*ic* acid is simply changed to -aldehyde:

Formic acid	Formaldehyde
(H-COOH)	(H-CHO)
Acetic acid	Acetaldehyde
(CH_3-COOH)	(CH_3-CHO)
Propionic acid	Propionaldehyde
(CH_3CH_2-COOH)	(CH_3CH_2-CHO)

FORMALDEHYDE

Formaldehyde is the first member of the aldehyde family, and it illustrates how precedent can be the predicate to principle. Formaldehyde's public image has always been associated with the funeral homes, doctor offices, and biology classes as an embalming fluid, a disinfectant, and a preservative. Even before 1900, it was produced by the oxidation of methanol. At the time methanol was known as wood alcohol since it came from the destructive distillation of wood. In 1905, Dr. Leo Baekeland in Yonkers, New York made a major breakthrough in the technology of plastics. He found a process for producing a stable, cross-linked polymer—later named Bakelite after him. The ingredients were phenol and formaldehyde. By the 1920s, the growth of this resin-strained wood alcohol producing capacity, but the revolutionary development of the methane reforming route to methanol relieved the situation. Despite the radical shift in methanol technology, the process for formaldehyde based on methanol feedstock has remained virtually unchanged even to today, despite the volume growth making it one of the top 25 commodity chemicals.

The process

The commercial process has always been to react methanol and air in the presence of a catalyst. Recent processes have switched from metal to metal oxide catalysts, especially iron oxide and molybdenum oxide.

$$CH_3OH + \tfrac{1}{2}O_2 \rightarrow H\text{-}CHO + H_2O$$

Methanol   Formaldehyde

The reactor in the formaldehyde plant, shown in Figure 16–1, is really just a heat exchanger—a large vessel that contains a bundle of tubes through which the methanol and oxygen are pumped. Outside the tubes, but inside the vessel, is a liquid that is used to transfer the heat. But the reaction is exothermic. So even though the reaction must take place at 575–700°F, once it starts, the exchanger must take heat away from the methanol + oxygen-to-formaldehyde reaction, rather than provide the heat to make it work. The heat-transfer liquid outside the tubes is continuously vaporized to take the heat away.

The tubes of the heat exchanger are filled loosely with the catalyst necessary for the reaction. Actual reaction time and residence in the tubes is less than one second.

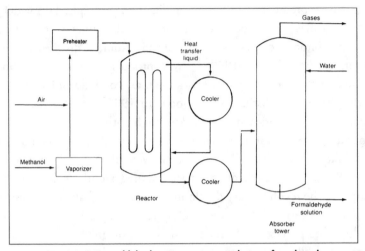

Fig. 16–1 Formaldehyde process—air oxidation of methanol

The hot reactor effluent gases are cooled to 230–265°F in a heat exchanger and passed into a water absorption tower. Formaldehyde is water-soluble and is separated from the remaining gases that exit the column overhead. Formaldehyde concentration in the tower is adjusted by controlling the amount of water added to the top of the tower. Generally, a product containing 37–56% formaldehyde in water is made. Methanol is often added as a stabilizer.

One of the major advantages of the metal oxide catalyst over that of the straight metal catalyst is the elimination of the need for a methanol recovery tower. The metal oxide catalysts result in not only high yields, but also very high conversion rates. Consequently, there is no need to recover the small amounts of methanol that remain unreacted. It becomes part of the aqueous formaldehyde solution and serves as a stabilizer for the system. By-products are CO, CO_2, dimethyl ether, and formic acid. The process yields (the percent of the methanol that ends up in formaldehyde) are 95–98%.

Material Balance

Feed:

Methanol	415 lbs.
Air (oygen)	209 lbs.

Product:

Formaldehyde	370 lbs.
Water	227 lbs.
By-products	33 lbs.

Properties and handling. Formaldehyde is a colorless, toxic gas at room temperature, with a pungent, irritating odor. It is flammable and explosive in presence of air. Both gaseous and liquid forms of formaldehyde polymerize at room temperature, and because of this, it can only be maintained in the pure state for a very short period. Because of these unhandy conditions, there are two ways formaldehyde gets into commerce, as a water solution called formalin and as a solid called paraformaldehyde or trioxane.

Formalin concentrations range from 37–56% by weight formaldehyde, the balance water, with or without stabilizer, but most formalin merchant sales is shipped as 50% or stronger solutions. The higher concentrations generally require stabilizers. No hazardous shipping label is required.

Formaldehyde Properties

Molecular weight	30.03
Freezing point	-180.4°F (-118°C)
Boiling point	-2.2°F (-19°C)
Specific gravity	0.815 (lighter than water)
Weight per gallon	6.8 lbs./gal.

Commercial aspects

More than 50% of the formaldehyde in the United States goes to making synthetic resins, primarily urea, melamine, and phenol-formaldehyde resins. These resins find multiple uses and applications as adhesives and insulation in housing construction and molded parts for the automotive, furniture, electronic, and appliance industries. Most of the rest of the formaldehyde is used as a solvent, adhesive intermediate, and printing inks component; and in the production of pentaerythritol (synthetic lubricants, varnishes), hexamethylene diamine (for Nylon 66), and pharmaceuticals. That leaves the smaller, older uses—fertilizer, germicide, disinfectant, embalming fluid, preservative. And since neatness and completeness count, you need to know that formaldehyde is used as a textile-sizing agent and accounts for the nice fresh look in your newer clothes.

Polymers of formaldehyde

Paraformaldehyde, $(CH_2O)_n$ where n is between 8 and 100, is a convenient polymer of formaldehyde. The polymer is easily formed by removing water from a 50% formalin solution under reduced pressure. As the formaldehyde concentration increases, crystals of paraformaldehyde form spontaneously. It is available at 91–97% purity. It is more stable than neat formaldehyde but just as useful in applications, where it readily decomposes back to the straight stuff.

Trioxane is a cyclic trimer (three formaldehydes in a ring). It is formed by distilling 56% formalin in the presence of sulfuric acid. At 99% purity, it has a melting point of 144°F and sublimes (goes directly to gases from the solid form) at 239°F. Trioxane is relatively stable and is soluble in water and the more common organic liquids. In the presence of a strong acid like hydrochloric or sulfuric, it easily depolymerizes to gaseous formaldehyde. This characteristic enables an immediate and controllable source of gaseous product for chemical reactions. No hazardous shipping labels are required for either grade.

```
                    Trioxane Properties
Molecular weight          90.5
Melting point             -145.0°F (63°C)
Boiling (subliming) point 239.0°F (115°C)
Specific gravity          1.39 (heavier than water)
Weight per gallon         11.7 lbs./gal.
```

ACETALDEHYDE

Acetaldehyde is old. It is not ancient like ethyl alcohol, the essential ingredient in wine, but it owes its discovery to this closely related compound. Scheele first prepared acetaldehyde in 1774 by dehydrogenation of ethyl alcohol. Just as many nicknames get attached to people at infancy, this process generated the name "aldehyde." It is a contraction for compounds that are alcohol dehydrogenates.

The close chemical relationship between acetaldehyde and ethyl alcohol is apparent in the grape fermentation process. The sugar in the grapes turns to acetaldehyde as an intermediate step. Fortunately for wine makers and oenophiles, the acetaldehyde immediately reduces to ethyl alcohol.

Oxidation of ethyl alcohol was one of the two important commercial routes to acetaldehyde until the 1950s. The other, much older route was the hydration of acetylene. The chemical industry was always after a replacement of acetylene chemistry, not just for acetaldehyde production, but all its many applications. Acetylene was expensive to produce, and with its reactive, explosive nature, it was difficult to handle. In the 1950s, acetylene chemistry and the ethyl alcohol oxidation route were largely phased out by the introduction of the liquid phase direct oxidation of ethylene. Almost all the acetaldehyde produced uses the newer process.

$$CH_2 = CH_2 + {}^1/_2O_2 \rightarrow CH_3\text{-}CHO$$

Ethylene $\quad$ Oxygen $\quad$ Acetaldehyde

Acetaldehyde also comes as a by-product in the production of vinyl acetate from ethylene and production of acrylic acid from propylene.

The process

The catalyst is the key to this reaction and in this case is an aqueous solution of palladium chloride ($PdCl_2$) and cupric chloride ($CuCl_2$). There is a complex, but well understood, mad scramble of ions and molecules that takes place as chlorine temporarily separates from the palladium and the copper and facilitates ethylene's reacting with oxygen.

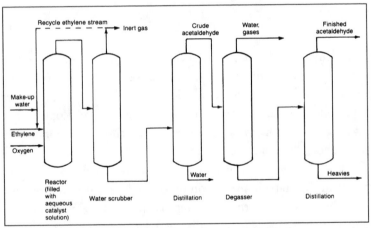

Fig. 16–2 Oxidation of ethylene to acetaldehyde

The process operates continuously and is easily regulated by the flow of fresh ethylene and oxygen to the reactor and the removal of acetaldehyde as a vapor.

As shown in Figure 16–2, high purity ethylene (99.7%) and oxygen (99.0%) are fed under pressure (100 psi) to a vertical reactor containing the aqueous catalyst solution. Reaction temperature is maintained at 250–275°F. Because the reaction is exothermic, heat liberated is partially removed by vaporizing the water present in the reactor. Makeup water is continuously fed to the reactor to maintain proper catalytic solution concentration.

The gaseous reaction mixture containing steam, unreacted ethylene, and acetaldehyde are passed into a water scrubber where the acetaldehyde is dissolved in water and then removed and fed to a distillation column. The overhead gases from the scrubber, ethylene, and water vapor are recycled to the reactor. Occasionally, the recycle gases must be purged to remove the build-up of inert by-product gases. In the distillation step, acetaldehyde is separated from water and a few by-product cats and dog. A final distillation step gives acetaldehyde in 99% purity.

Material Balance

Feed:

Ethylene	670 lbs.
Oxygen	404 lbs.
Catalyst	small

Product:

Acetaldehyde	1000 lbs.
By-products	4 lbs.

Other processes

A few plants have been built to oxidize normal paraffins such as propane and butane. Air and paraffin are charged to a tubular furnace at a temperature of about 700°F. Acetaldehyde yields from butane are about 30–35%.

Methanol can be hydroformylated with syngas (CO and H_2) in the presence of a copper or nickel halide catalyst to give acetaldehyde. Reaction conditions are 350–400°F and 4000–6000 psi.

Commercial aspects

Acetaldehyde is a colorless, flammable liquid with a pungent, fruity-like odor and has a boiling point close to room temperature (70°F). It is soluble in water and most common organic solvents. It is a toxic chemical requiring care in handling. Acetaldehyde is commercially available as technical grade, 99% minimum purity. Because of its low boiling temperature, acetaldehyde must be contained in pressure vessels, including rail cars and tank trucks, which during a shipment fly the hazardous material placard.

Acetaldehyde Properties

Molecular weight	44.05
Freezing point	-190.3°F (-123.5°C)
Boiling point	69.6°F (20.9°C)
Specific gravity	0.779 (lighter than water)
Weight per gallon	6.49 lbs./gal.

As early as World War I, acetaldehyde was the primary route to acetic acid and acetone. While other preferred technologies for acetone have been developed, acetaldehyde remains an important intermediate to acetic acid as well as several other chemicals.

Product	End Uses
Acetic acid and anhydride	Vinyl acetate, textile processing, cellulose acetate (cigarette filters) and aspirin
Pentaerythritol	Alkyd resins, explosives, synthetic lubricants, coatings
C₄ alcohols	Solvent, plasticizer, chemical intermediate
2-ethyl hexyl alcohol	Plasticizer for PVC

These four derivatives of acetaldehyde account for more than 80% of its total U.S. production.

Polymers of acetaldehyde

Like formaldehyde, acetaldehyde easily forms polymers, in this case paraldehyde and metaldehyde. Paraldehyde will form when hydrochloric or sulfuric acid is added to acetaldehyde. Polymerization of acetaldehyde to metaldehyde occurs in the gaseous phase in the presence of aluminum oxide or silicon dioxide catalyst.

Paraldehyde is a colorless liquid, hazardous, flammable, miscible with most organic solvents, and soluble in water. If you leave it alone, it will slowly decompose back to acetaldehyde, so it is often used as a substitute for acetaldehyde. It is also used as a rubber accelerator, an antioxidant, and a solvent for fats and oils.

Metaldehyde is a white solid that will also decompose partially to acetaldehyde, but only if it is heated above 175°F. It is hazardous, flammable, and of little commercial use.

Paraldehyde Properties	
Molecular weight	132.16
Boiling point	255.9°F (124.4°C)
Melting point	54.5°F (12.5°C)
Specific gravity	0.992
Weight per gallon	8.33 lbs./gal.

Chapter 16 in a nutshell ...

Aldehydes have the characteristic -CHO group, sometimes written as

$$-C\overset{\displaystyle O}{\overset{\displaystyle \|}{H}}$$

-CH. They are a dehydrogenated form of a corresponding alcohol. Formaldehyde, CH_2O, corresponds to methanol, acetaldehyde, CH_3CHO, to ethyl alcohol.

The formaldehyde process is an air oxidation of methanol, CH_3OH, which has water as a by-product. Formaldehyde is a gas at room temperature, but is usually handled either as a water solution called formalin or as polymers called paraformaldehyde and trioxane. Both are readily converted back to formaldehyde. Some uses of formaldehyde are the manufacture of polymer resins and as a germicide.

Acetaldehyde is made by the direct oxidation of ethylene, C_2H_4. It is a liquid at room temperature and is an intermediate in the production of acetic acid, acetic anhydride, butyl, and 2-ethyl hexyl alcohol.

EXERCISES

1. Fill in the blanks.

    a. An aqueous solution of formaldehyde:_____

    b. Will polymerize:_____

    c. Trioxane is a trimer of _____that easily reverts to
    _____.

    d. A chemical intermediate:_____

    e. Product of complete oxidation:_____

    f. Dehydrogenation of alcohol:_____

    g. Feedstock for acetic acid:_____

2. What is an oenophile anyway, and why do they have an aversion to acetaldehyde?

3. Write out the oxidation chain that shows the progression from ethane to CO_2 + H_2O and includes the appropriate olefin, alcohol, aldehyde, and acid.

CHAPTER 17

The Ketones
Acetone, Methyl Ethyl Ketone, and Methyl Isobutyl Ketone

"Names are not always what they seem.
The common Welsh name —Bzjxxllwcp—
is pronounced Jackson."

Mark Twain, 1835–1910

There's no need for ill-humored comments about the ketones being a trio of pop singers. They're a family of organic compounds (first cousins to the aldehydes) that all have the ketone signature

$$- C -$$
$$\|$$
$$O$$

Somewhere in the middle of a hydrocarbon chain, a double-bonded oxygen replaces the two hydrogens attached to a carbon. Ketones come in many sizes and shapes. The convention for naming them is to refer to the alkyl groups attached to the ketone signature. In Figure 17–1, the three commercially traded aliphatic ketones with the largest volumes are shown: acetone, MEK, and MIBK.

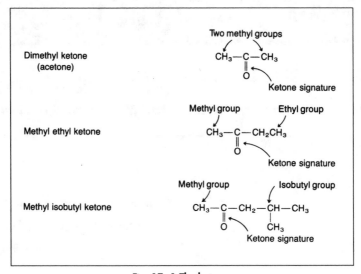

Fig. 17–1 The ketones

ACETONE

Right up front, you need to know that acetone and dimethyl ketone (DMK) are the same thing chemically.

As far back as the pre-World War I years, acetone was used extensively as a solvent. The early commercial routes to producing acetone included the destructive distillation of wood, the fermentation of either starch or corn syrup, and the conversion of acetic acid. The development of olefin technology permitted a more efficient petrochemical route, the dehydrogenation of isopropyl alcohol, to replace the originals by the late 1930s. It remained the primary route through the 1960s.

One convenient way to classify today's processes for making acetone is to separate them into two categories, by-product and on-purpose. You'll unquestionably recall that acetone is one of the outturns of the cumene-to-phenol process described in Chapter 7. (Approximately 0.6 pounds of acetone are generated for each pound of phenol.) That falls into the category of by-product production because the rate at which acetone is produced is not solely dependent on anticipated acetone demand. Often the demand for the phenol dictates the rates at which the phenol plant is run, and either the acetone is a disposal problem or a shortage exists. More than 90% of the acetone produced in the United States is in the "by-product" category.

The "swing" supply of acetone comes from the plants that produce acetone "on-purpose" by catalytic dehydrogenation of IPA. There is nearly as much capacity in place in the United States to produce on-purpose acetone as there is via the phenol route. But on-purpose being the swing supply, the growing demand for phenol has resulted in by-product acetone producers shutting down most of the on-purpose capacity.

The process

Like a lot of other petrochemical processes, the "chemistry" part of the plant, the reactor, is simple. Together with all the other "mechanical" processes like heating, cooling, and especially separation that fill up the plant site, these make the acetone plant look like a typical petrochemical plant, if there is such a thing.

The dehydrogenation route is shown in Figure 17–2. In this plant, the isopropyl alcohol feed is heated to about 900°F in a preheater and then charged to a reactor at about 40–50 psi pressure. The reactor is filled with a catalyst of zinc oxide deposited on pumice. Pumice is a fine powder of silica dioxide, i.e., glass. It has many fine pores in which the catalyst can reside, and therefore has a very large surface area to expose to the IPA. The catalyst causes the hydrogen to pop off the -OH group, forcing the double bond to the oxygen, the ketone signature.

The hot effluent from the reactor containing acetone, unreacted IPA, and hydrogen is cooled in a condenser and then scrubbed with water to remove the hydrogen. Both IPA and acetone are highly soluble in water, but hydrogen is not. So, by washing the effluent with water, the hydrogen bubbles out the top, and the IPA/acetone comes out the bottom with the water.

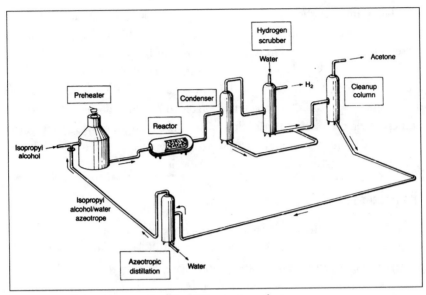

Fig. 17–2 IPA-to-acetone plant

The process is just like the solvent recovery scheme explained in Chapter 2. The IPA/acetone/water is fractionated with the acetone going overhead and the IPA/water coming out the bottom. The yield (the percent of the IPA that ends up as acetone) is 85–90%.

The presence of the water causes special treatment of the IPA recycle stream. IPA and water will form an azeotropic solution, like the one discussed in Chapter 3. The two compounds will boil together (in specific proportions) at a temperature different than the boiling point of either. Consequently, the stream recycled to the reactor contains about 9–10% water. The presence of water doesn't affect the IPA dehydrogenation step. It's just a little extra baggage, requiring a little more heating, cooling, and pumping around.

Material Balance

Feed:

Isopropyl alcohol	1158 lbs.

Product:

Acetone	1000 lbs.
H_2	34 lbs.
By-products	124 lbs.

Other routes

The more expedient, direct catalytic oxidation route to acetone was developed in Germany in the 1960s. If you had been in charge of building the acetone business from scratch, you'd probably not have built any IPA-to-acetone plants if you had known about the Wacker process. It's a catalytic oxidation of propylene at 200–250°F and 125–200 psi over palladium chloride with a cupric (copper) chloride promoter. The yields are 91–94%. The hardware for the Wacker process is probably less than for the combined IPA/acetone plants. But once the latter plants were built, the economies of the Wacker process were not sufficient to shut them down and start all over. So the new technology never took hold in the United States.

There are several other routes to acetone of minor importance: air oxidation of IPA; reaction between IPA and acrolein for the production of allyl alcohol, with acetone as the by-product; vapor phase oxidation of butane; coproduction when IPA is oxidized yielding acetone and H_2O_2, hydrogen peroxide, the principal ingredient of bleach; and by-product production from the manufacture of methyl ethyl ketone.

Commercial aspects

Uses. U.S. acetone production is used primarily in two basically different ways, 70% of it as a chemical intermediate and 20% as a solvent. As an intermediate, acetone is used to produce MIBK, methyl methacrylate (used to make Plexiglas products), Bisphenol A (raw material for epoxy and poly-carbonate resins), and higher molecular weight glycols and alcohols.

As a solvent, acetone is used in varnishes, lacquer, cellulose acetate fiber, cellulose nitrate (an explosive), and as a carrier solvent for acetylene in cylinders. Acetylene is stored at about 225 psi but is so explosively reactive that as an extra precaution the cylinder is filled with asbestos wool soaked in acetone. Acetylene is extremely soluble in acetone, and the asbestos keeps it from sloshing around when the cylinder is half empty. Acetone also is used in smaller volumes for the manufacture of pharmaceuticals and chloroform (the anesthetic).

Properties and handling. Acetone is a mobile, colorless, volatile, highly flammable liquid. It has an odor that makes you think you're in a hospital. Acetone dissolves in water, alcohol, ether, and most other organic solvents. That's why it's usually included in paintbrush cleaner. It dissolves almost anything and then can be washed away with water.

Acetone Properties	
Molecular weight	58.08
Freezing point	-139.6 °F (-95.4 °C)
Boiling point	133.0 °F (56.1 °C)
Specific gravity	0.7901 (lighter than water)
Weight per gallon	6.6 lbs./gal.

Acetone is sold commercially in three grades, USP (99%), CP (99.5%), and technical (99.5%). The terms USP and CP are acronyms used in the trade and stand for *U.S. pure* and *chemical pure*. Acetone is shipped in run-of-the-mill tank trucks, in tank cars, and in drums. The hazardous material shipping placard must be displayed for this highly flammable liquid.

METHYL ETHYL KETONE

Most of what you read in the previous section about acetone also applies to methyl ethyl ketone (MEK). The processes for making MEK can be broadly categorized into "by-product" and "on-purpose"; the more popular processes are the same—they just start with larger molecules, and the applications are much the same.

The three popular manufacturing routes to MEK are:

1. catalytic dehydrogenation of secondary butyl alcohol (instead of IPA)
2. air oxidation of butylene (instead of propylene)
3. catalytic oxidation of butane to form acetic acid and by-product MEK

In the first route, not only is there close similarity in the chemistry of the acetone/dehydrogenation route, the hardware is almost identical to the plant shown in Figure 17–2. The reaction is as follows:

$$
\begin{array}{cc}
\text{OH} & \text{O} \\
| & || \\
CH_3 - CH\text{ - }CH_2 - CH_3 \;\rightarrow\; CH_3 - C - CH_2 - CH_3 \;+\; H_2 \\
\text{(SBA)} & \text{(MEK)}
\end{array}
$$

The heated secondary butyl alcohol (SBA) vapors are passed through a reactor containing zinc oxide catalyst at 750–1000°F at atmospheric pressure. The catalyst causes the hydrogen to pop off, forming MEK. The separation of the reactor effluent into MEK, water, hydrogen, and recycled SBA is about the same as Figure 17–2. The overall yield (the percent of SBA that ends up as MEK) is about 85–90%.

Material Balance

Feed:

Secondary butyl alcohol   1140 lbs.

Product:

Methyl ethyl ketone	1000 lbs.
Hydrogen	28 lbs.
By-products	112 lbs

A more energy-efficient version of this process takes place in the liquid phase. A catalyst of very fine Raney nickel or copper chromate, suspended in a heavy, high boiling temperature solvent, is mixed with SBA. At 300–325°F the SBA undergoes dehydrogenation to MEK. As it does, the MEK and hydrogen immediately vaporize, leaving the reaction medium in gaseous form and need only to be separated from each other.

The second, on-purpose route to MEK is the direct oxidation of butylene, the Wacker process:

$$2CH_2 = CH - CH_2 - CH_3 + O_2 \rightarrow 2CH_3 - \overset{\displaystyle O}{\overset{\displaystyle \|}{C}} - CH_2 - CH_3$$
$$\text{Butylene} \hspace{6cm} \text{MEK}$$

With reaction conditions of 200–225°F, 150–225 psi, and a palladium chloride-cupric chloride catalyst, MEK yields are 80–90%. The operating costs of the Wacker process for MEK (and acetone and several other petrochemicals as well) are relatively low. But the plant is made of more expensive materials. Because of the corrosive nature of the catalyst solution, critical vessels, and the piping are titanium-based (that's expensive!), and the reactor is rubber-lined, acid-resistant brick.

The third route, catalytic oxidation of butane, producing by-product MEK, accounts for only a modest portion of the total supply, less than 15%. Plants designed to produce acetic acid from the direct oxidation of butane can be run to produce almost no MEK. But optimum operating cost balanced against market product prices usually warrants shifting to a 60/40 acetic acid/MEK outturn.

Commercial aspects

MEK is used in a variety of ways as a solvent. It owes part of its popularity to the fact that it is a low boiling point replacement for the butyl alcohols in vinyl, nitrocellulose, acrylic, and other coatings. In these applications, MEK flashes (vaporizes or "dries") quickly at ambient temperatures, leaving the coating behind. However, MEK is classified as a hazardous material, putting special requirements on where and how it can be used.

MEK also is used as the solvent in lube oil dewaxing, wood pulping, and toluene (*see* Chapter 3), and in the manufacture of printing ink and rubber-based industrial cements.

Properties and handling. The physical characteristics of MEK are similar to those of acetone. It's colorless, mobile, flammable, and sweet-smelling, if that's what you call a hospital smell. It's very soluble in water and most common organic solvents. There are only two grades commercially traded, technical (99%), and CP (99.95%). Shipping and handling are similar to acetone.

Methyl Ethyl Ketone Properties	
Molecular weight	72.1
Freezing point	-123.5 °F (-86.4 °C)
Boiling point	175.3 °F (79.6 °C)
Specific gravity	0.806 (lighter than water)
Weight per gallon	6.7 lbs./gal.

METHYL ISOBUTYL KETONE

MIBK is more complicated than the one-step conversion process for acetone and MEK. Manufacture of MIBK takes the three-step process shown in Figure 17–3, starting with acetone.

First, the acetone is condensed (or reacted or dimerized) with itself. That is, it's passed over a catalyst and two acetone molecules chemically react to form diacetone alcohol. Diacetone alcohol has both the acetone signature and the alcohol signature (-OH). The catalyst is an alkaline compound like $Ca(OH)_2$ (calcium hydroxide or soda lime), and the reaction is run at about 32°F.

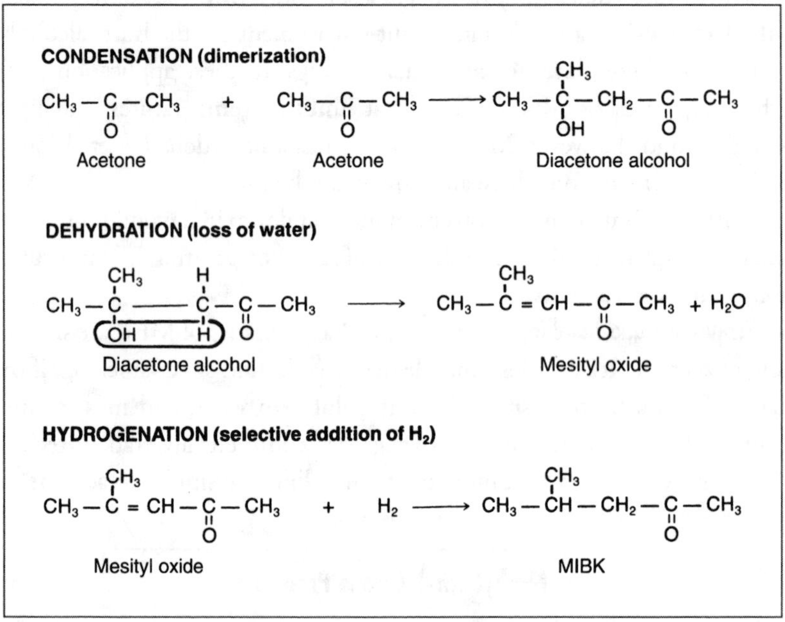

Fig. 17–3 MIBK process reactions

In the second step, the diacetone alcohol is dehydrated (the -OH group and a hydrogen atom are clipped off) to form mesityl oxide. The dehydration is done by mixing the diacetone alcohol with the water-loving catalyst sulfuric acid at 212–250°F.

In the third step, the mesityl oxide is hydrogenated (hydrogen added) to MIBK by heating it to the vapor stage at 300–400°F and passing it over a copper or nickel catalyst at 50–150 psi in the presence of hydrogen.

Material Balance

Feed:

Acetone	1160 lbs.
Hydrogen	20 lbs.

Product:

Methyl isobutyl ketone	1000 lbs.
Water	138 lbs.
By-products	44 lbs

One problem with this process is the difficulty of controlling the last step. As the MIBK gets formed, it also has the tendency to hydrogenate further to methyl isobutyl carbinol. Further addition of hydrogen wipes out the ketone signature, replacing it with the hydroxyl group, -OH. This unavoidable by-product, methyl isobutyl carbinol, has to be separated from the MIBK by fractionation. The overall yield of MIBK (the amount of acetone that ends up as MIBK) is around 90%.

Alternate processes

A small amount of MIBK is made from a new European-originated process. A complex catalyst system involving palladium metal and a cation exchange resin is used. The reaction permits going directly from acetone to MIBK.

A process similar in concept involves going directly from IPA to a mixture of acetone and MIBK. The process is confidential, and the producers are good at keeping secrets so no details are available at this writing.

Commercial aspects

Uses. The applications of MIBK read a lot like those of MEK. In the 1960s and 1970s, MIBK rapidly replaced the use of ethyl acetate and butyl acetate as a solvent for resins. However, MIBK is now classified as a hazardous material, limiting its applications as a coating material and solvent to special situations or enclosed systems.

Some unique applications for MIBK include metallurgical extraction (particularly plutonium from uranium), a reaction solvent in pharmaceuticals, an adhesive, and, if you stretch the definition of applications, the manufacture of methyl isobutyl carbinol.

Properties and handling. MIBK is a colorless liquid and has a pleasant, almost fruity, odor. Unlike acetone and MEK, it is only slightly soluble in water. That happens to solvents as the size of the molecule gets larger. Most of the commercial trade in MIBK is in the technical grade (98.5%). Bulk shipments can be handled in conventional tank trucks and tank cars, but the hazardous material markings must be displayed.

MIBK Properties	
Molecular weight	100.16
Freezing point	-119.0 °F (-84.0 °C)
Boiling point	241.0 °F (116.0 °C)
Specific gravity	0.8024 (lighter than water)
Weight per gallon	6.7lbs./gal.

Chapter 17 in a nutshell ...

$$O$$
$$\|$$

Ketones have the characteristic -C- signature group imbedded in them. Acetone, CH_3COCH_3, comes from two different routes. It is a by-product in the cumene to phenol/acetone process. It is the "on-purpose" product of the catalytic dehydrogenation of isopropyl alcohol. Acetone is popular as a solvent and as a chemical intermediate for the manufacture of MIBK, methyl methacrylate, and Bisphenol A.

Methyl ethyl ketone (MEK) and methyl isobutyl ketone (MIBK) have higher boiling temperatures, are less hazardous liquids, and are also popular as solvents. MEK is made by dehydrogenation of secondary butyl alcohol or the direct oxidation of butylene. MIBK is made via a three-step process starting with acetone.

EXERCISES

1. Why are aldehydes and ketones "first cousins"? What are the signature groups of the two?

2. MIBK competes in the coatings solvent market with ethyl acetate. Because of its superior properties, MIBK can sell at a premium. When ethyl acetate is selling at 40 cents/lb., MIBK can fetch 45 cents/lb.
   The Skim Chemical Company can rent some IPA plant, acetone plant, and MIBK plant capacity from the Takesits Toll Company for 5 cents/lb. of product. When ethyl acetate is 40 cents/lb., what is the breakeven price that Skim can pay for propylene if the by-products of each plant are worthless but hydrogen costs 20 cents/lb.?

3. Fill in the blanks from the list on the next page.

   a. _____ Used as an anesthetic.
   b. _____ Feedstock for MIBK.
   c. _____ Will polymerize.
   d. _____ Used as a solvent.
   e. _____ Hydrogenation catalyst.
   f. _____ Dehydrogenation catalyst.
   g. _____ Feedstock for acetone.
   h. _____ Direct oxidation of propylene.
   i. _____ Major chemical use for acetone.
   j. _____ Feedstock for MEK.
   k. _____ Lube oil dewaxing agent.
   l. _____ By-product in MIBK production.
   m. _____ Oxidation catalyst.

isopropyl

acetone

chloroform

MIBK

methyl methacrylate

palladium chloride

ether

zinc oxide

nickel

propylene

MEK

methyl isobutyl carbinol

Wacker process

sesame seeds

ethylene

formaldehyde

butylene

secondary butyl alcohol

CHAPTER 18 ⬡

The Acids

> "Eye of a newt and toe of a frog,
> Wool of a bat and tongue of a dog."
>
> ***MacBeth***
> **Shakespeare, 1564–1616**

There are dozens of organic acids that are used in petrochemicals processing. But there are three that account for more than 70% of the total volume—acetic acid, adipic acid, and the phthalic acids. These compounds have little in common with each other besides the carboxyl signature group, written as -COOH, drawn as

$$-C\text{-}OH$$
$$\parallel$$
$$O$$

The group is so-called because it's a combination of the carbonyl (-C=O) and hydroxyl (-OH) groups.

You can think of carboxylic acids as being third down the line on the route to oxidizing a paraffin completely.

Paraffin	ethane	$CH_3\text{-}CH_3$
Alcohol	ethyl alcohol	$CH_3\text{-}CH_2\text{-}OH$
Aldehyde	acetaldehyde	$\begin{array}{c} O \\ \parallel \\ CH_3\text{-}C\text{-}H \end{array}$
Acid	acetic acid	$\begin{array}{c} O \\ \parallel \\ CH_3\text{-}C\text{-}OH \end{array}$
	carbon dioxide and water	$CO_2 + H_2O$

Organic acids can be aliphatic or aromatic, and can be mono-, di-, or polycarboxylic. Aliphatic acids have paraffinic hydrocarbon chains as their roots. The higher molecular weight aliphatic acids, the ones with greater than 12 carbons, are often referred to as the fatty acids because many of them were originally obtained by the hydrolysis of animal fat or vegetable oil. The word aliphatic is from the Greek *aleiphatos*, meaning fat.

The aromatic acids, as you would suspect from the name, have a benzene ring connected directly to the carboxyl signature group. Dicarboxylic acids have carboxyl groups attached in two places. Monocarboxylic acids have only one, and of course, the poly acids have three or more.

ACETIC ACID

Acetic acid is one of the simplest members of the aliphatic acid family. It has a methyl group attached to the acid signature group: CH_3COOH. Acetic acid is easily the largest volume organic acid produced in the United States (more than 3 billion pounds per year). There are several places you're likely to find acetic acid or its derivatives. Acetic acid is the natural component of vinegar that gives it the characteristic smell. (*Acetum* is the Latin word for vinegar.) Acetic acid also is used to make acetates, which are polymerized and processed into adhesives and water base paints (polyvinyl acetate) and fibers (cellulose acetate, like Arnel).

Acetates are esters of acetic acid. Remember from Chapter 1 that an ester has a signature group

$$-\underset{\underset{O}{\|}}{C}-O-R$$

The ester's name usually comes from the acid and ends in the suffix *-ate*, as does the word acetate itself. The acetate group is made from acetic acid by replacing the carboxyl hydrogen with an *R* group:

$$CH_3-\underset{\underset{O}{\|}}{C}-O-R.$$

If the *R* is the vinyl group $-CH=CH_2$, for example, you have vinyl acetate,

$$CH_3-\underset{\underset{O}{\|}}{C}-O-CH = CH_2.$$

If the *R* is the ethyl group $-C_2H_5$, then you have ethyl acetate,

$$CH_3-\underset{\underset{O}{\|}}{C}-O-C_2H_5.$$

Manufacturing acetic acid

The destructive distillation of wood to produce methanol results in some by-product acetic acid, and that was the most popular but now defunct commercial source. Fermentation, the oldest, indeed the ancient method, is still used to produce vinegar for the food industry. Vinegar is a 3–5% solution of acetic acid in water.

Most of the "on-purpose" acetic acid is made by one of the following routes:

Oxidation of Acetaldehyde
$$CH_3CHO + \tfrac{1}{2}O_2 \rightarrow CH_3COOH$$

Oxidation of Butane
$$CH_3CH_2CH_2CH_3 + O_2 \rightarrow CH_3COOH + \text{by-products}$$

Carbonylation of Methanol
$$CH_3OH + CO \rightarrow CH_3COOH$$

Table 18–1 summarizes the vital statistics of the various routes. The methanol route is the latest in state-of-the-art technology and is now the preferred route to acetic acid. More than 75% of the acetic acid in the United States is from carbonylation of methanol.

Process	Catalyst	Reaction °F	Pressure psi	Yield %	By-Product
Butane oxidation	Cobalt acetate	300–450	800	57	Acetaldehyde + acetone + methanol
Acetaldehyde oxidation	Manganese acetate	150	0	95	None
Methanol carbonylation	Rhodium iodide	350–475	200	99	None

Table 18–1 Acetic acid processes

Acetic acid plants

Acetaldehyde process. A stainless steel, water jacketed kettle is filled with concentrated (99%) acetaldehyde and the manganese acetate catalyst. Then air is bubbled through for about 12 hours, where it reacts with the acetaldehyde. The gases exiting the kettle, still mostly air, are bubbled through water to "scrub" them and then discharged to the atmosphere. Some of the acetaldehyde ends up in the scrubbing water, but this is recovered by distillation.

The mixture in the kettle, which is crude acetic acid, is distilled to glacial acetic acid, 99% purity.

Butane process. Butane is added to a vessel that contains a solution of acetic acid and the cobalt acetate catalyst. The pressure is kept high enough to keep the butane liquid. Air is then bubbled through at 300–450°F and mixed vigorously. The volatile by-products come out the top—methane, carbon dioxide, and unreacted air. Crude acetic acid is drawn off the bottom and distilled to give glacial acetic acid. Yield is 75–80%. The word *glacial* indicates the high purity form of acetic acid, 99.5–99.8%, as opposed to plain old commercial grades of acetic acid that are diluted with water from 99.8% all the way down to 3% acetic.

Methanol process. BASF introduced high-pressure technology way back in 1960 to make acetic acid out of methanol and carbon monoxide instead of ethylene. Monsanto subsequently improved the process by catalysis, using an iodide-promoted rhodium catalyst. This permits operations at much lower pressures and temperatures. The methanol and carbon monoxide, of course, come from a synthesis gas plant.

The reaction is run at about 350°F and 200 psi and yields about 99% acetic acid. That is, about 99% of the methanol that gets converted ends up as acetic acid. By-products include only small amounts of dimethyl ether and methyl acetate.

Conversion rates as high as 99% are not encountered very often in the petrochemical industry. That coupled with relatively mild operating conditions, made this route the economic favorite since it was introduced. About 75% of the world's acetic acid production comes from the methanol route.

Material Balance

Feed:

Methanol	540 lbs.
Carbon monoxide	473 lbs.
Catalyst	small

Product:

Acetic acid	1000 lbs.
By-products	13 lbs.

BP Chemical reports success using an iridium-based catalyst that gives higher reaction rates and less by-product formation, though when you already have 99% conversion, "less" is relatively small. The main advantage of the BP experience seems to be lower operating costs because of the speedier reaction, saving energy and capital.

Commercial aspects

More than 65% of the acetic acid produced in the United States goes into vinyl acetate. Nearly all the vinyl acetate ends up as polyvinyl acetate, used to make plastics, latex paints, and adhesives. About 12% of acetic acid is converted to acetic anhydride that is mostly used to make cellulose acetate, the white stuff in cigarette filters. It is also used in the manufacture of plastic sheeting and film and in formulating lacquers.

Acetic acid also finds use as a chemical intermediate in the production of acetate esters for paint solvents and as a reaction solvent for the manufacture of terephthalic acid. Also, acetic acid is the source of the acetyl group in the manufacture of acetyl salacylic acid.

Acetic Acid Properties	
Molecular weight	60.05
Boiling point	244.6 °F (118.1 °C)
Freezing point	61.9 °F (16.6 °C)
Specific gravity	1.0492 (heavier than water)
Weight per gallon	8.64 lbs./gal.

Properties and handling. Acetic acid has the strong, pungent odor of vinegar. It is a colorless liquid that is soluble in water and most organic solvents. The concentrations of the commercial grades vary all over, as low as 3%. The USP glacial acetic acid is 99.8% pure.

Since acetic melts at 62°F, shipping pure grades poses a special problem. Cool weather can cause freezing, expansion, and container rupture.

Tank cars and trucks must be specially lined because of the reactive nature of the acid. Even dilute acetic acid will react if left long enough, as any chef who has made sauerbraten will attest. That's how he got that tough piece of chuck steak so tender—by reacting (soaking) it in vinegar, among other things. Culinary considerations aside, the white hazardous (corrosive) shipping label is required for concentrated acetic acid.

ADIPIC ACID

You can use analogies to put adipic acid in its right place. Acetic acid is the most important aliphatic monocarboxylic acid; adipic is the most important aliphatic dicarboxylic acid. (You remember, of course, that carboxylic is the contraction for carbonyl and hydroxyl, -C=O and -OH, or together, -COOH. Right?) Also, adipic acid is to Nylon 66 what cumene is to phenol. About 95% of the adipic acid ends up as Nylon 66, which is used for tire cord, fibers, and engineering plastics.

Adipic acid is produced by oxidizing cyclohexane. The two-step process shown in Figure 18–1 is used for almost all production. Cyclohexane is oxidized with air over a cobalt naphthenate catalyst to give a mixture of cyclohexanol and cyclohexanone. These two products are separated from the unreacted cyclohexane and then hit with a 50% nitric acid solution. That opens up the C_6 ring and adipic acid is formed. Yields are in the 90–95% range.

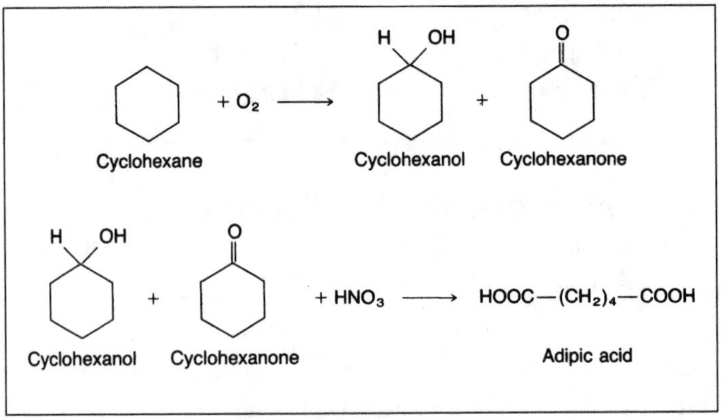

Fig. 18–1 Adipic acid process

An alternate route to get cyclohexanone is sometimes used—hydrogenating phenol in the liquid phase catalyzed by palladium on carbon. The rest of the process is then the same, except that yields are in the 70% range. Since the by-product yield is so high, the process has had limited acceptance, with only about 5% of the adipic acid being made this way.

Material Balance

Feed:

Cyclohexane	625 lbs.
Nitric acid	excess
Air	470 lbs.
Catalyst	small

Product:

Adipic acid	1000 lbs.
NO + NO$_2$	small
N$_2$O, N$_2$, and air	residual
By-products	77 lbs.

Commercial aspects

Since almost all adipic acid is used for Nylon 66, the primary producers are Nylon 66 manufacturers.

The fibers made from Nylon 66 are durable, tough, and abrasion-resistant, which suits them for tire cord. They are easy to color, which gives them a secure place in the carpet market (and on the floor). The additional attributes of moldability or processibility make Nylon 66 suitable in the engineering plastics market.

Properties and handling. Adipic acid doesn't physically fit the usual image of an acid. Its melting temperature is 306°F. At normal temperatures, it is a white, crystalline powder that can be transported in one-ton cartons and in drums and 50-pound bags. Adipic acid is only slightly soluble in water but dissolves in alcohol. The commercially traded grade is 99.5% pure.

Adipic acid is an approved food additive and is one of the few solid petrochemicals manufactured on a commercial scale. (Terephthalic acid is another.)

PHTHALIC ACIDS

The strange spelling of these acids comes from the shortening of the original form, naphthalic acids. Naphtha originally came from an ancient Iranian word that was pronounced neft. It referred to a flammable liquid that oozed out of the earth. The word later anglicized to naphtha. Lexicography aside, the phthalic acids are made from the three xylenes, orth-, meta-, and para-xylene, and are shown in Figure 18–2.

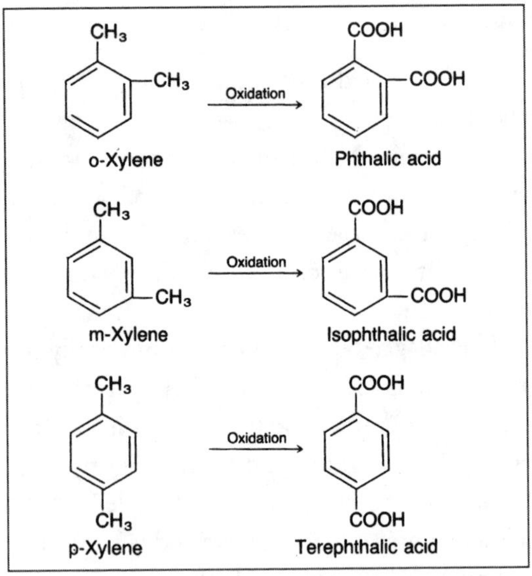

Fig. 18–2 Xylenes and phthalic acids

The major uses of these dicarboxylic acids (two -COOH groups in each) are plasticizers for polymers, alkyd and polyester resins, and fibers. All these applications are discussed in Chapters 22–24.

Phthalic acid and phthalic anhydride

In the primary application of phthalic acid, life is rather transitory. Almost all phthalic acid is used to make phthalic anhydride. When ortho-xylene is used as the starting base chemical, phthalic acid is formed but immediately dehydrates (loses a molecule of water) to form phthalic anhydride, as shown in Figure 18–3.

Until 1959, all the phthalic anhydride was made from coal tar naphthalene, the double-benzene ring compound also shown in Figure 18–3 was easily oxidized directly to phthalic acid. But with phthalic anhydride being only a small share of coal oil, and with the demand for phthalic anhydride escalating rapidly, coal tar became an inadequate source. The frantic search for an alternative route led to the development of the recovery process for ortho-xylene from refinery aromatics streams discussed in Chapter 3 and the conversion of ortho-xylene to phthalic acid and anhydride. With the continued growth in the need for plasticizers and the inelasticity of naphthalene supply, ortho-xylene now accounts for 90% of the phthalic anhydride supply in the United States.

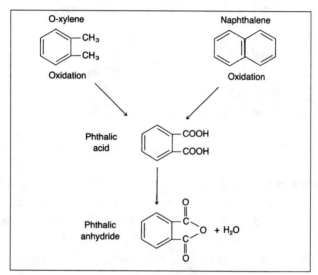

Fig. 18–3 Phthalic anhydride processes

The process. A typical process for phthalic anhydride starts with mixing hot o-xylene vapor with excess preheated air in a ratio of 20:1 air to o-xylene by weight. The gaseous mixture is then fed to a reactor consisting of tubes packed with vanadium pentoxide catalyst on a silica gel. The reaction takes place at 800–1000°F. Like most oxidation reactions, this one is exothermic, and the heat of reaction must be removed from the tubes to maintain the reaction temperature.

Contact time between the reactants and the catalyst is about a tenth of a second. The reaction gases—mainly phthalic anhydride, carbon dioxide, and water—are cooled, condensed, and purified in stainless steel facilities. Phthalic anhydride solidifies at 269°F, so the purified (99.5%) product can be stored in its molten form or cooled and flaked. Minor amounts of by-products, maleic anhydride, phthalic acid, and benzoic acid are also produced.

Material Balance

Feed:

Ortho-xylene	975 lbs.
Excess air (7 times)	206 lbs. (oxygen)

Product:

Phthalic anhydride	1000 lbs.
Water	497 lbs.
By-products	210 lbs.
Unreacted oxygen	1329 lbs.

Properties and handling. At ambient temperatures, phthalic anhydride is a white crystalline solid. It is slightly soluble in water. It is commercially available in two grades—pure (99.5%) and technical (99%). It is shipped in drums and bags in the solid form. Liquid phthalic anhydride is shipped in heated tank cars and trucks. It is not classified as a hazardous material because it is not corrosive or flammable.

Phthalic Anhydride Properties

Molecular weight	148.1
Melting point	269.0 °F (131.6 °C)
Boiling point	563.2 °F (295.1 °C)
Specific gravity	1.527(heavier than water)
Physical appearance	White crystalline flakes or needles

Applications. Phthalic anhydride is used largely to make plasticizer for polyvinyl chloride. It is also a feed for alkyd resins and for unsaturated polyesters that are widely used in construction, marine, and synthetic marble applications. Other minor applications are dyes, esters, drying oil modifiers and pharmaceutical intermediates.

Terephthalic acid

The sole use for para-xylene is to make terephthalic acid (TPA) and its derivative, dimethyl terephthalate (DMT). When DMT is copolymerized with ethylene glycol, chemists call it polyethylene terephthalate. On Seventh Avenue in New York, they call it polyester. On the labels, it is sometimes called Dacron.

The acid, TPA, is not unstable like phthalic acid. TPA can't dehydrate to the anhydride because the two acid groups, -COOH, aren't in the right places. So for the most part, TPA is the product that is traded commercially and producers make DMT where they make the polyester.

The original route from p-xylene was oxidation in the presence of nitric acid. But the use of nitric acid is always problematical. There are corrosion and potential explosion problems, problems of nitrogen contamination of the product, and problems due to the requirement to run the reactions at high temperatures. Just a lot of problems that all led to the development of the liquid air phase oxidation of p-xylene. Ironically the nitrogen contamination problem was the reason that the intermediate DMT route to polyester was developed, since that was easy to purify by distillation. Subsequently, DMT has secured a firm place in the processing scheme.

High purity DMT is produced by esterification of TPA. That is, the terephthalic acid is reacted with methanol to form the ester (actually a di-ester), in Figure 18-4.

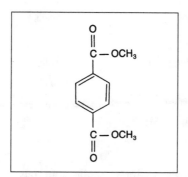

Fig. 18–4 Dimethyl terephthalate

In the DMT process, the esterification is done by feeding a slurry of TPA crystals in methanol to a reactor with a catalyst of sulfuric acid at 220°F and 50 psi. DMT forms and can be purified by distillation. Yields exceed 95%, based on the TPA that ends up as DMT. In some later designs resulting in less severe operating conditions, MEK or acetaldehyde have been used as promoters in place of sodium bromide.

The TPA process. The technology involves the oxidation of p-xylene, as shown already in Figure 18–2. The reaction takes place in the liquid phase in an acetic acid solvent at 400°F and 200 psi, with a cobalt acetate/manganese acetate catalyst and sodium bromide promoter. Excess air is present to ensure the p-xylene is fully oxidized and to minimize by-products. The reaction time is about one hour. Yields are 90–95% based on the amount of p-xylene that ends up as TPA. Solid TPA has only limited solubility in acetic acid, so happily the TPA crystals drop out of solution as they form. They are continuously removed by filtration of a slipstream from the bottom of the reactor. The crude TPA is purified by aqueous methanol extraction that gives 99+% pure flakes.

Material Balance	
Feed:	
Para-xylene	680 lbs.
Air (excess)	1843 lbs.(oxygen)
Catalyst	small
Product:	
Terephthalic acid	1000 lbs.
Water	230 lbs.
By-products	53 lbs.
Unreacted oxygen	1240 lbs.

Properties and handling. Terephthalic acid has a high melting point, 572°F. At room temperature, it is a white crystalline solid, insoluble in water or acetic acid. It is commercially available in fiber grade (99%) and technical grade (96%). Just to confuse things, the fiber grade of TPA is referred to as

PTA. TPA and PTA are routinely shipped in bags, drums, and hopper cars as flakes. No hazardous shipping label is required.

Terephthalic Acid Properties

Molecular weight	166.14
Melting point	Sublimes* at 572.0 °F (300.0 °C)
Specific gravity	1.51 (heavier than water)
Physical appearance	White crystals or powder

*Goes directly from solid to vapor without passing through the liquid phase.

Applications. About 95% of the TPA is used to make polyester. Most of that goes into fiber production, some into films (magnetic tapes, photographic materials, and electrical insulation). The route to fibers is through DMT or through a direct process if PTA is used. Minor amounts of TPA are used for herbicides, adhesives, printing inks, coatings, and paints. Polybutylene terephthalate is a molding resin used as an engineering plastic.

Isophthalic acid

The stepsister of the other two phthalic acids is isophthalic acid made from meta-xylene. The applications are similar, but the commercial demand is smaller. If it weren't for the fact that m-xylene is a coproduct of the other xylenes, no one would invent it. The other xylenes and phthalic acids would probably suffice.

Isophthalic acid is made by the same process as TPA, liquid phase air oxidation. Yields are about 80%. Isophthalic does have some unique redeeming value—it will enhance, to some extent, the mechanical and temperature sensitive properties of polyesters, alkyd resins, and glass reinforced plastics.

Chapter 18 in a nutshell ...

Organic acids can be thought of as oxidation of corresponding alcohols, since they have the characteristic -OH signature group, plus a double-bonded oxygen: -COOH or

$$
\begin{array}{c}
\text{O} \\
\parallel \\
\text{–C-OH.}
\end{array}
$$

Acetic acid, CH_3COOH, can be made by the oxidation of acetaldehyde, CH_3CHO; by catalytic addition of CO to methanol; or by butane oxidation. Most acetic acid is used to make vinyl acetate or cellulose acetate, which are the intermediates for plastics, paints, adhesives, yarn, and cigarette filters.

Adipic acid is made by the reaction of nitric acid and cyclohexane. The adipic acid has two -COOH groups, which makes it very reactive. Adipic is used primarily for making Nylon 66.

The phthalic acids are made by oxidation of the corresponding xylene isomer. They are used to make plasticizers and alkyd and polyester resins and fibers. Ortho-phthalic acid usually is not isolated because it loses a molecule of water so easily, forming phthalic anhydride, the commercially traded form of this strain of phthalic acid.

EXERCISES

1. What is the relationship of an acid to:

   a. an ester?
   b. an aldehyde?
   c. an acid anhydride?
   d. carbon dioxide?

2. How do you distinguish an acid from an alcohol?

3. What chemical property does acetic acid and phthalic acid have in common?

4. Adipic acid is to _____ as cumene is to _____ as ethylene dichloride is to _____ as naphthalene is to _____ as ethylbenzene is to _____.

5. Insert in the middle column the intermediate compound structure, and identify in all three columns the name of the compounds:

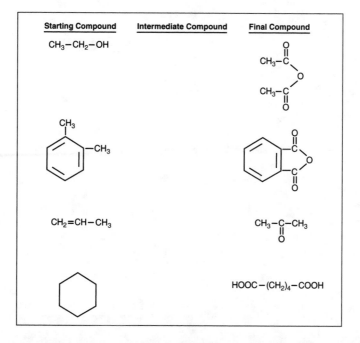

CHAPTER 19 ◇

Acrylonitrile, Acrylic Acid, and the Acrylates

"I am a Bear of Very Little Brain,
and long words Bother me."

Winnie the Pooh
A. A. Milne, 1882-1956

A crylonitrile, acrylic acid, and acrylates…they all sound alike; the molecules in Figure 19–1 look similar, they are all covered in this same chapter. So they must be well connected. Right?

Well, not quite. First, they sound similar because at one time, one of them, acrylonitrile, was based solely on manufacture from acrolein, a pungent liquid whose roots in Latin are *acer,* meaning sharp, and *olere,* meaning smell. Acrylonitrile was made from acrolein, and acrylates were derivatives of acrylonitrile. But acrylates also are made from acrylic acid, which is also a derivative of acrylonitrile. So the name, acrylo, covers an extended family of relations.

Second, the similarity in structure has nothing to do with the relationship between the two. The double bonded group on the left of each molecule in Figure 19–1 is the vinyl group. What gives each of them their interesting and unique properties are the acid group

$$\text{(-C-OH)}$$
with =O above

interesting and unique properties are the acid group (-C-OH) in acrylic

acid, the ester group (-C-OR) in acrylates, and the nitrile or cyanide group (-CN) in acrylonitrile. Indeed, acrylonitrile used to be called vinyl cyanide, but that was before the petrochemical industry had good public relations people.

$$CH_2 = CH - CN$$
Acrylonitrile

$$CH_2 = CH - \overset{\overset{O}{\|}}{C} - OH$$
Acrylic acid

$$CH_2 = CH - \overset{\overset{O}{\|}}{C} - OR$$
Acrylates

Fig. 19–1 Acrylonitrile, acrylic acid, and the acrylates

By now you should be wondering where this chapter will take you. So Figure 19–2 is the road map. You can see right away why the relationships are a little confusing. Acrylates can be made from acetylene, propylene, acrylonitrile, or acrylic acid. Acrylic acid can be made from propylene or acrylonitrile. And acrylonitrile can be made from propylene, but it also is used to make other, unrelated but important things too, namely polymers or adiponitrile, the precursor to Nylon 66. If you keep this road map handy, struggling through the next few pages will be easier.

Along the way there is one important processing fact to keep in mind. Acrylonitrile or acrylic acid may be an intermediate step to acrylates, but sometimes the intermediate is not isolated (separated or recovered) as a commercial product. That is what makes it difficult to separate discussion of the three chemicals in a neat, orderly way.

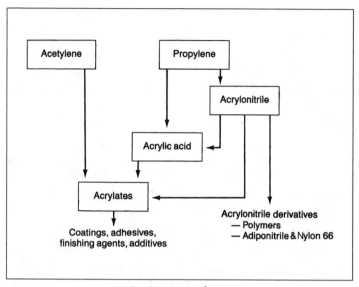

Fig. 19–2 A road map

Acrylonitrile

The nitriles are a group of compounds that can be thought of as derivatives of hydrogen cyanide, HCN. The hydrogen is removed and replaced by an organic grouping. In the case of acrylonitrile, the replacement is the vinyl grouping, $CH_2=CH-$, the same one encountered in styrene and vinyl chloride.

The original route to acrylonitrile was the catalytic reaction of HCN with acetylene. That was a combination of two compounds that together had all the characteristics you'd like to avoid—poisonous, explosive, corrosive, and on and on. But during World War II, acrylonitrile became very important as a comonomer for synthetic rubber (nitrile rubber). Later, the growth for acrylonitrile came from synthetic fibers like Orlon, Acrylon, and Dynel.

In the 1960s, like almost all acetylene technology, the HCN/C_2H_2 route to acrylonitrile gave way to ammoxidation of propylene. That word, ammoxidation, looks suspiciously like the contraction of two more familiar terms, ammonia and oxidation, and it is. When Standard of Ohio (Sohio) was still a company they developed a one-step vapor phase catalytic reaction of propylene with ammonia and air to give acrylonitrile.

$$CH_2 = CH\text{-}CH_3 + NH_3 + {}^{3}/{}_{2}O_2 \rightarrow CH_2 = CH\text{-}CN + 3H_2O$$

As a by-product, HCN is also formed, but there generally is a ready market for it. In fact, this process has become an important commercial source of HCN.

The plant

The early ammoxidation plants were a two-step design. Propylene was catalytically oxidized to acrolein (CH_2=CHCHO). The acrolein was then reacted with ammonia and air at high temperature to give acrylonitrile. The one-step process has replaced most of this hardware.

The Sohio technology is based on a catalyst of bismuth and molybdenum oxides. Subsequent catalyst improvements came from the use of bismuth phosphomolybdate on a silica gel, and more recently, antimony-uranium oxides. Each change in catalyst was motivated by a higher conversion rate per pass to acrylonitrile.

The propylene stream, shown in Figure 19–3, can be either refinery grade (50–70% propylene) or chemical grade (90–95%). The propylene, ammonia, and oxygen are fed in a ratio of 1:1:2 to the vessel containing the catalyst. The vessel is called a fluidized bed reactor because the catalyst moves about like a fluid. The catalyst is usually a very fine, hard powder that flows very easily. As the reactants pass through the vessel, they are mixed with the catalyst. Because the catalyst particles are so small and there are so many, the total surface area of the catalyst exposed to the gaseous or liquid reactants is huge. So the yields from fluidized bed reactors are generally higher than fixed bed reactors. The main disadvantage is the loss of catalyst because of the difficulty of mechanically separating the particles out after the reaction is complete.

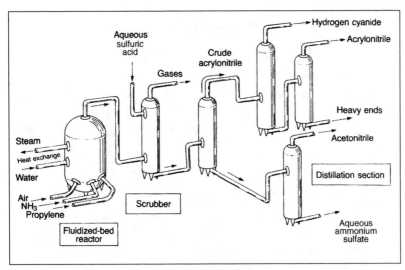

Fig. 19–3 Acrylonitrile plant

The ammoxidation reaction is carried out at about 800°F and 30 psi. Because it is highly exothermic, heat is removed continuously from the reactor by heat exchangers. The residence time of the reactants is about three seconds. The reaction gases are cooled as they pass by the water-to-steam heat exchanger in the reactor. The effluent is treated for removal of ammonia by scrubbing it with water acidified with sulfuric acid, forming ammonium sulfate, a marketable commodity that can be recovered by crystallization. But that's another story.

The gases (unreacted propylene, CO_2, and air) are removed in an absorption column. The rest of the products go with the aqueous solution to be separated in a series of fractionating columns. The major by-products are water, CO_2, acetonitrile, and hydrogen cyanide.

The major difficulties with these processes are controlling heat removal from the reactor; the stability of the catalyst, both mechanical and chemical; and catalyst loss. The latter two problems are due to the use of the fluidized bed reactor. Yields of acrylonitrile from this process are about 70%, based on propylene feed.

Material Balance	
Feed:	
Propylene	1175 lbs.
Ammonia	475 lbs.
Air (excess O_2)	1631 lbs. O_2
Catalyst	Small
Product:	
Acrylonitrile	1000 lbs.
Unreacted O_2 + NH_3 approx.	819 lbs.
HCN	100 lbs.
Water	1008 lbs.
By-products	354 lbs.

Other processes

Just as in other areas, researchers are fiddling with catalyst to find something that will enable them to use a paraffin instead of an olefin as the starting point, in this case, propane which is cheaper than propylene 99% of the time. Several catalysts have been developed that can be used, but the reaction temperatures are higher (about 950°F) and the residence times a lot longer (about 15 seconds), so existing hardware generally cannot handle the new technology. The yields, based on propane, are about 70%, the same as with propylene, but until the operating requirements improve, there won't be enough incentive to build new plants to shut down the existing ones.

Commercial aspects

Uses. Acrylic fibers account for about half the acrylonitrile production. Orlon, Acrylon, and Dynel are polymers and copolymers of acrylonitrile. These fibers find extensive usage in apparel and household furnishings as well as in the industrial markets.

Nitrile rubber has declined in importance, but has been replaced by styrene-acrylonitrile (SAN) copolymers and acrylonitrile-butadiene-styrene

(ABS) terpolymers. These plastics are relatively inexpensive, tough, and durable, and they were the first so-called engineering plastics to capture sizable pipe and auto parts markets.

A more recent use of acrylonitrile is its use to make adiponitrile, which is the feedstock used in Nylon 66 production. Acrylonitrile also has been found to be good treatment for cotton, making it resistant to mildew, heat, and abrasion, and more receptive to dyes.

Properties and handling. Acrylonitrile is a colorless, flammable liquid with a boiling point of 171°F. It is traded commercially as technical grade (99%) and is bulk shipped in lined tank cars or trucks with the hazardous material markings. The linings are necessary due to the corrosive nature of acrylonitrile.

Acrylonitrile Properties	
Molecular weight	53.1
Freeze point	-117.0 °F (-83.0 °C)
Boiling point	171.1 °F (77.3 °C)
Specific gravity	0.811 (lighter than water)
Weight per gallon	6.74 lbs./gal.

Methacrylonitrile

Methacrylonitrile can be produced in the acrylonitrile plants by ammoxidation of isobutylene, which gives it the methyl group sticking out.

$$CH_3$$
$$|$$

This molecule, $CH_2 = C - CN$, is copolymerized with acrylic acid, styrene, maleic anhydride, butadiene, or isoprene to produce a wide variety of plastics and coatings.

ACRYLIC ACIDS

Many petrochemicals have been harnessed because they have two common characteristics: they're simple and they're reactive. Acrylic acid (AA) is the simplest organic acid that contains a double bond. It's that vinyl group again, $CH_2=CH-$, the same one found in acrylonitrile, styrene, and vinyl chloride. Because it's an acid and because it has the double bond, it's highly reactive. It readily undergoes polymerization (reacts with itself because of the double bond) and esterification (reacts with alcohol because it's an acid).

The use of acrylic acid can be traced at least as far back as about 1900. It was an additive for paints and lacquers. Due to the tendency for acrylic acid to polymerize at low temperatures, it accelerated the "drying" process. The users probably didn't understand the chemistry of polymerization at the time, only that it worked.

Early routes to AA were complex and expensive. In 1927 the ethylene chlorohydrin process was introduced, but it was also still expensive, and not much commercial interest was stimulated in AA. In 1940 a process came literally right off the farm—pyrolysis of lactic acid, a waste product of the dairy industry found in sour milk.

$$\begin{array}{cc} \underset{\substack{|\\ \\ |\\ \\}}{\overset{\substack{|\\ \\ |\\ \\}}{H-C-C-COOH}} & \\ \end{array}$$

$$\underset{\text{(Lactic acid)}}{\overset{\displaystyle H \;\; H}{\underset{\displaystyle H \;\; OH}{\overset{|\quad\;|}{\underset{|\quad\;|}{H-\,C\!-\!C\!-\!COOH}}}}} \;\;\rightarrow\;\; \underset{\text{(Acrylic acid)}}{\overset{\displaystyle H}{\overset{|}{CH_2\!=\!C\!-\!COOH}}} \;+\; H_2O$$

This route improved the economics of AA some, because of the availability of zero-cost raw material, the lactic acid. But the operating costs were still too high for rapid commercialization. It wasn't until the 1950s, with the Reppe process route to acrylic acid, starting with acetylene, and then in the 1960s, starting with propylene, did acrylics begin to take off.

With that meandering, historical background, it's better to switch to the subject of acrylates to deal with the specific manufacturing routes. In some, the acrylic acid, though formed in the process, never gets recovered as a commercial project. It just forms then converts to an acrylate.

Acrylic Acid Properties	
Molecular weight	72.06
Freeze point	53.8 °F (12.1 °C)
Boiling point	285.6 °F (140.9 °C)
Weight per gallon	8.84 lbs./gal.

ACRYLATES (AND METHACRYLATES)

These are fun to read about because they end up in all sorts of products you're familiar with, but probably never thought too much about. To begin with, you need to understand what acrylates are.

If you take an alcohol (a compound with an -OH signature) and react it with an organic acid (one with a -COOH signature), the product is an ester (the -COOR signature) and the process is called esterification. If the organic acid you use is acrylic acid, the ester is called an acrylate. And if the alcohol is, say, methyl alcohol, then the product is methyl acrylate, but not methacrylate. If you start out with methacrylic acid, then you get a methacrylate. And finally, if you use methyl alcohol and methacrylic acid, you get methyl methacrylate, which is a big star in petrochemicals.

You'll recall (of course) from Chapter 1 that the letter "R" is used as a substitute for a carbon group, like methyl, ethyl, etc. The general equation for esterification of acrylic acid is:

$$\underset{\text{Acrylic acid}}{CH_2{=}CH{-}\overset{\displaystyle O}{\overset{\displaystyle \|}{C}}{-}OH} + \underset{\text{Alcohol}}{ROH} \rightarrow \underset{\text{Acrylate}}{CH_2{=}CH{-}\overset{\displaystyle O}{\overset{\displaystyle \|}{C}}{-}OR} + \underset{\text{Water}}{H_2O}$$

Specifically, for the reaction with methyl alcohol,

$$CH_2 = CH - \overset{\overset{\displaystyle O}{\displaystyle \|}}{C} - OH + CH_3OH \rightarrow CH_2 = CH - \overset{\overset{\displaystyle O}{\displaystyle \|}}{C} - OCH_3 + H_2O$$
Methyl acrylate

The major commercial acrylates are formed from the alcohols that should by now be familiar to you—methanol, ethanol, butanol, iso-butanol, and 2-ethyl hexanol. The corresponding acrylates are shown in Table 19–1.

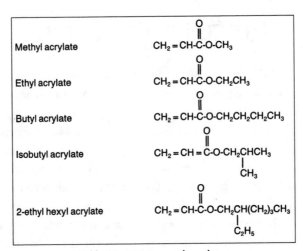

Table 19–1 Commercial acrylates

Acetylene to ethyl acrylate

The Reppe process was commercialized in the 1950s. It involves the reaction of acetylene, carbon monoxide, and an alcohol (methyl, ethyl, etc.) to give an acrylic ester (an acrylate). The process is carried out at 125°F and 15–30 psi in a nickel carbonyl/aqueous hydrochloric acid solution. The nickel carbonyl acts as both a catalyst and a secondary source of carbon monoxide.

$$HC\equiv CH + CO + CH_3\text{--}CH_2\text{--}OH \rightarrow CH_2\text{=}CH\text{--}\overset{\overset{\displaystyle O}{\|}}{C}\text{--} OCH_2\text{--}CH_2$$

Acetylene   Ethanol   Ethyl acrylate

Carbon monoxide

AA also can be made from this method by leaving out the alcohol and modifying the operating conditions. The conventional esterification reaction to produce the acrylates can then be run. The lower molecular weight acrylates (methyl and ethyl) are usually produced via the "direct" technology. The higher molecular weight acrylates are usually made from methyl or ethyl acrylate by what chemists call transesterification reaction. The higher weight alcohol does a little square dance with the acrylate, changing partners by replacing the methyl or ethyl group with a higher weight group such as a butyl or 2-ethyl hexyl group.

$$CH_2\text{=}CH\text{--}\overset{\overset{\displaystyle O}{\|}}{C}\text{--}O\text{--}CH_3 + C_4H_9OH \rightarrow CH_2\text{=}CH\text{--}\overset{\overset{\displaystyle O}{\|}}{C}\text{--}O\text{--} C_4H_9 + CH_3OH$$

Methyl acrylate   n-Butanol   Butyl acrylate   Methanol

Reppe process yields are about 80%, but the usual acetylene drawbacks are present: hazardous materials handling and higher-cost raw materials. As a result the acetylene route plants are not being duplicated as they wear out, giving way to the newer technologies.

Hydrolysis of acrylonitrile

The cost of producing acrylonitrile dropped when the ammoxidation process was introduced in the 1960s. Then it became economical at that time to produce methyl and ethyl esters of acrylic acid by hydrolyzing acrylonitrile in the presence of alcohol. The hydrolysis and esterification take place at the same time, in the presence of sulfuric acid at about 225°F. Yields are about 98%.

$$CH_2=CH-CN + 2H_2O + H_2SO_4 \rightarrow CH_2=CH-\overset{\overset{O}{\|}}{C}-OH + (NH_4)HSO_4$$

Acrylonitrile   Acrylic acid   Ammonium bisulfate

$$CH_2=CH-\overset{\overset{O}{\|}}{C}-OH + CH_3OH \rightarrow CH_2=CH-C-OCH_3 + H_2O$$

Acrylic acid   Methanol   Methyl acrylate

The process consumes the sulfuric acid and produces a waste product, ammonium bisulfate, so it is expensive. So when propylene oxidation technology was developed, it became the preferred route.

Material Balance

Hydrolysis of Acrylonitrile:
Feed:

Acrylonitrile	541 lbs.
Ethyl alcohol	469 lbs.
Water	184 lbs.
Sulfuric acid	980 lbs.

Product:

Ethyl acrylate	1000 lbs.
Ammonium bisulfate	1150 lbs.
By-products	24 lbs.

Esterification of Acrylic Acid:
Feed:

Acrylic acid	846 lbs.
Methyl alcohol	376 lbs.
Sulfuric acid catalyst	trace

Product:

Methyl acrylate	1000 lbs.
Water	211 lbs.
By-products	11 lbs.

Catalytic oxidation of propylene

The newest and most commercially successful process involves vapor phase oxidation of propylene to AA followed by esterification to the acrylate of your choice. Chemical grade propylene (90–95% purity) is premixed with steam and oxygen and then reacted at 650–700°F and 60–70 psi over a molybdate-cobalt or nickel metal oxide catalyst on a silica support to give acrolein (CH_2=CH-CHO), an intermediate oxidation product on the way to AA. Other catalysts based on cobalt-molybdenum vanadium oxides are sometimes used for the acrolein oxidation step.

Acrolein is immediately passed through a second oxidation reactor to form acrylic acid. The reaction takes place at 475–575°F, over a tin-antimony oxide catalyst. A few by-products form, namely, formic acid (HCOOH), acetic acid (CH_3COOH), low molecular weight polymers, carbon monoxide, and dioxide. But overall yields of propylene to acrylic acid are high—85 to 90%.

Material Balance

Propylene Oxidation
Feed:

Propylene	642 lbs.
Oxygen	735 lbs.
Catalyst	Small

Product:

Acrylic acid	1000 lbs.
Water	250 lbs.
By-products	127 lbs.

Oxidative Carbonylation of Ethylene

A route not yet commercialized is the reaction of ethylene, carbon monoxide, and air to give AA. The ethylene is dissolved in acetic acid. The process takes place at 270°F and 1100 psi in the presence of palladium chloride-copper chloride catalyst. Yields are 80–85%. If the by-product and corrosion problems can be licked, the process will probably catch on.

Commercial aspects

Uses. The most commercially important acrylates are ethyl-, butyl-, 2 ethyl hexyl-, and methyl-acrylate, in that order. Major markets include surface coatings, adhesives, leather and textile finishing agent, paper coating, and cement additives. An important feature of the acrylates is that they readily polymerize if exposed to heat, light, oxygen, or peroxides. Most important, they polymerize in water to form a latex, which is a dispersion of solid particles in water, such as latex paints. A little diversion here might give a better understanding of the value of acrylates.

Emulsion polymerization was developed as part of the synthetic rubber program during World War II. Take an acrylate monomer (an unpolymerized acrylate) and add it to water. It's immiscible—doesn't mix. If you add an emulsifying agent like soap (yes, soap), the acrylate becomes dispersible (miscible) in water. Now add a little water soluble catalyst. That induces polymerization. The acrylate monomer links itself chemically to other acrylate monomers and, as the polymer molecules grow to the right weight and size, they can be stabilized. The resulting mixture is called a latex. Add color pigment, and you've got the basics for a latex paint.

Acrylic latices (more than one latex) find many uses in the field of coatings. Every amateur house painter appreciates the handling advantages:

1. When exposed to air and light, the latex will further polymerize to a hard coating at a moderate speed. (It "dries" fast.)

2. Before it polymerizes ("sets up"), it is soluble in water. (Easy cleanup of brushes and painter.)

3. After it polymerizes, it is stable and resists oxidation. (It's weather resistant and colorfast.)

4. During the drying process, only water vaporizes. With oil-based paints, naphtha or mineral spirits vaporize during drying. (Latex paints don't pollute the atmosphere.)

Handling. Acrylates are traded as technical grade (99% purity), inhibited or uninhibited. Usually they are sold with trace amounts of hydroquinone as an inhibitor.

Methyl and ethyl acrylates are toxic enough to require a hazardous shipping label, but butyl-, isobutyl-, and 2-ethyl hexyl-acrylates have high enough flash points to be considered safe.

Ethyl Acrylate Properties	
Molecular weight	100.06
Freeze point	98.0 °F (-72.0 °C)
Boiling point	211.3 °F (99.6 °C)
Specific gravity	0.923 (lighter than water)
Weight per gallon	7.68 lbs./gal.

METHACRYLATES

The methacrylates are first cousins to the acrylates, but only one member of this branch of the family ever made it into commercial big-time, methyl methacrylate (MMA). The most important feature of MMA is that it polymerizes into a transparent or translucent plastic.

$$
\begin{array}{c}
O \\
\parallel \\
CH_2{=}C{-}C{-}OCH_3 \\
\mid \\
CH_3
\end{array}
$$

Methyl methacrylate

Process

The production of MMA has long been accomplished by the old standby acetone cyanohydrin route. (*See* Figure 19–4.) Acetone reacts with hydrogen cyanide in the presence of an aqueous solution of sodium hydroxide at 100–150°F to give acetone cyanohydrin. The MMA is then produced by hydrolyzing acetone cyanohydrin in the presence of 98% sulfuric acid and methyl alcohol. The two-step reaction occurs at about 200°F. After purification, overall yield is 80–85%.

Material Balance

Feed:

Acetone	581 lbs.
Hydrogen cyanide	270 lbs.
Methanol	320 lbs.
Sulfuric acid (98%)	981 lbs.

Product:

Methyl methacrylate	1000 lbs.
Ammonium bisulfate	1152 lbs.

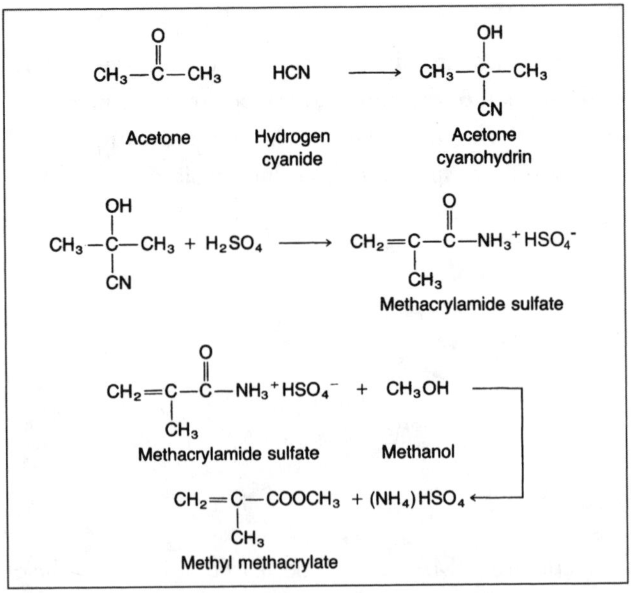

Fig. 19–4 Methyl methacrylate synthesis

Several alternate routes to MMA eliminate ammonium bisulfate by-product—really a coproduct since it's 1.5 tons for every 10 tons of MMA produced. They also do not involve HCN, always a safety problem in the plants and sometimes an unreliable market. Although these routes are more efficient and economical, American producers have stuck to the acetone cyanohydrin route. The plants are fully amortized and by staying with the old technology, producers can avoid the large capital investments associated with a new plant.

In Asia, Asahi and Mitsubishi have commercialized a process using isobutylene or tertiary butyl alcohol to make methacrolein. Then they further oxidize it to methacrylic acid, MAA, which is then esterified with methanol to MMA. The same process might eventually start with isobutane oxidation to bypass the olefin step.

BASF led the development of a route based on ethylene and synthesis gas. Its four step process begins with the production of propionaldehyde from ethylene, CO, and H_2 using a proprietary catalyst mixture that they aren't telling anything about. Reaction with formaldehyde gives methacrolein. The last two steps are the same as above—oxidation with air yields the MAA; subsequent reaction with methanol yields MMA.

1) $CH_2{=}CH_2 + CO + H_2 \rightarrow CH_3{-}CH_2{-}CHO$
      Ethylene   Syngas   Propionaldehyde

2)

$$CH_3{-}CH_2{-}CHO + CH_2O \rightarrow \underset{\text{Methacrolein}}{CH_2{=}\overset{\overset{\displaystyle CH_3}{\displaystyle |}}{C}{-}CHO} + H_2O$$
                                        Formaldehyde

3)

$$\underset{\text{Methacrolein}}{CH_2{=}\overset{\overset{\displaystyle CH_3}{\displaystyle |}}{C}{-}CHO} + {}^1\!/_2 O_2 \rightarrow \underset{\text{Methacrylic acid}}{CH_2 = \overset{\overset{\displaystyle CH_3}{\displaystyle |}}{C}{-}COOH}$$

4)
$$CH_2=\underset{\underset{MAA}{\underset{|}{CH_3}}}{C}-COOH + CH_3OH \rightarrow CH_2=\underset{\underset{MMA}{\underset{|}{CH_3}}}{C}-COOCH_3$$

Commercial aspects

Uses. The sole commercial use of MMA is polymers in various forms—cast sheets, latices, and molding and extrusion polymers. MMA homopolymers (polymers that use only one monomer, as opposed to copolymers or terpolymers that use two or three) are best known for their use in the form of clear, transparent sheets with trade names like Plexiglas and Lucite. Applications include advertising signs, aircraft windows, desktops, lighting fixtures, building panels, and plumbing and bathroom fixtures.

MMA is also used extensively as a copolymer with acrylates in latex paints and as a homopolymer in lacquers, since it's transparent.

MMA molding and extrusion polymers are used in the automotive industry for control dials, knobs, instrument covers, directional light covers, and tailgate lenses. The last two are probably the largest application of MMA molding powders.

A is also used in conjunction with other plastics to achieve translucent or transparent qualities. Transparent bottles, made by copolymerization of MMA with vinyl chloride, are gradually replacing glass containers. MMA is used in many of the same applications as acrylic latices and is also used as a comonomer with acrylonitrile to make acrylic fibers.

Methyl Methacrylate Properties	
Molecular weight	100.1
Freeze point	-54.8 °F (-48.2 °C)
Boiling point	212.2 °F (100.1 °C)
Specific gravity	0.938 (lighter than water)
Weight per gallon	7.86 lbs./gal.

Properties and handling. MMA is a colorless, sweet smelling, volatile liquid that boils at 212°F. MMA readily polymerizes with itself, and usually has trace amounts of hydroquinone added as an inhibitor. MMA is traded as technical grade and is shipped in lined tank cars, tank trucks, and drums. The hazardous material warnings are required on all shipments.

Chapter 19 in a nutshell ...

Acrylonitrile, C_2H_3CN or $CH_2=CH-CN$, has the characteristic nitrile signature group, CN. The double bond between the carbons makes "acrylo" useful in polymerizations as an intermediate in the manufacture of acrylates and adiponitrile for Nylon 66 production. The primary route to acrylo is the reaction of ammonia and oxygen with propylene. The poor match of atoms in and out results in only 70% yield.

Acrylic acid, $CH_2=CHCOOH$, has the characteristic acid signature group. Acrylic acid can be made from propylene or from acrylonitrile and is generally used to make acrylates.

The acrylates (for example, ethyl acrylate, $CH_2=CH-COOCH_2CH_3$) are esters of acrylic acid, so they end in the suffix -*ate* and have the characteristic signature -COOR.

The methacrylates, which are commercially even more important than the acrylates, are esters of methacrylic acid and are used extensively in coatings, plastics, and adhesives.

EXERCISES

1. What feedstock do acrylonitrile and acrylic acid have in common? How about acrylates and acrylic acid?

2. What are the major differences between a fixed bed catalyst process and a fluidized bed process? What are the advantages of a fluidized bed process? The disadvantages?

3. The vinyl grouping, $CH_2=CH-$, imparts chemical reactivity to an organic compound. List a half dozen or so compounds containing this reactive grouping.

4. Esterification and dehydration reactions both produce water. How do they differ?

5. Would "olefins" be considered suitable feedstock for manufacturing acrylonitrile, propyl acrylate, MMA, and methacrylonitrile? Elaborate.

CHAPTER 20 ⬡

Maleic Anhydride

*"How many apples fell on Newton's head
before he took the hint?!"*

Robert Frost, 1874–1963

The unlikely molecule in Figure 20–1 is a cyclic anhydride known by several names: 2-butene-1,4-dicarboxylic acid anhydride; cis-butene-dioic acid anhydride; maleic anhydride (MA); and when you've been in the business a long time, "maleic."

The name, maleic anhydride, came about in the same fashion as any number of compounds early in the petrochemical business. Many organic acids and their derivatives were given common names based on some early observations, their special source in nature, or on some special feature of their structure. MA was first isolated in the 1850–75 era by dehydration of malic acid, a sugar acid found in apple juice. The Latin word for apple is malum. Hence, malum, malic, maleic. The suffix, anhydride, which follows each alias of MA, has a simple definition: a compound derived by the loss of a molecule of water from two carboxyl groups (-COOH).

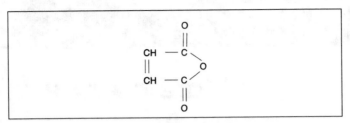

Fig. 20–1 Maleic anhydride

You may wonder why a chemical with such an unusual structure is so popular. The answer, as always, is reactivity. But in this case, MA is thrice blessed. If you refer to Figure 20–1, there is reactivity associated with:

- the anhydride group on the right
- the double bond on the left
- the carboxylic acid grouping that (re-)forms when MA is mixed with water (That's the group that gave up the water molecule in the previous paragraph that formed MA to begin with.)

MA can be produced from several different feedstocks—benzene, normal butenes, and normal butane. The popularity of any one of them has swayed with the economic winds that set the feedstock prices. Benzene was the original choice in the 1940–50 period. Butene came on strong in the 1950s and faded quickly as they became surplus in refineries and then short again. The divergence between the prices for normal butane and butene or benzene has stimulated interest in butane. Moreover, normal butane has a yield advantage. If you study the MA molecule for a moment, you'll see that it has only four carbons. When benzene (C_6H_6) is the feedstock, two carbon atoms must be eliminated to form MA. They end up as a waste product, CO_2. When butane (C_4H_{10}) is the feedstock, the atoms eliminated are the lightweights, hydrogen. The theoretical yields of MA are 1.26 pounds of MA per pound of benzene but 1.75 pounds of MA per pound of normal butane. So the yield advantage is no small economic factor.

The benzene and butane routes are very similar. Benzene route hardware is often adaptable to the butane route because the pressures, temperatures, and even the catalyst are the same. For that reason, benzene plants have been converted to butane plants fairly cheaply. Today, practically all the MA produced in the United States is based on butane feed. Elsewhere in the world, the favorite feed is benzene.

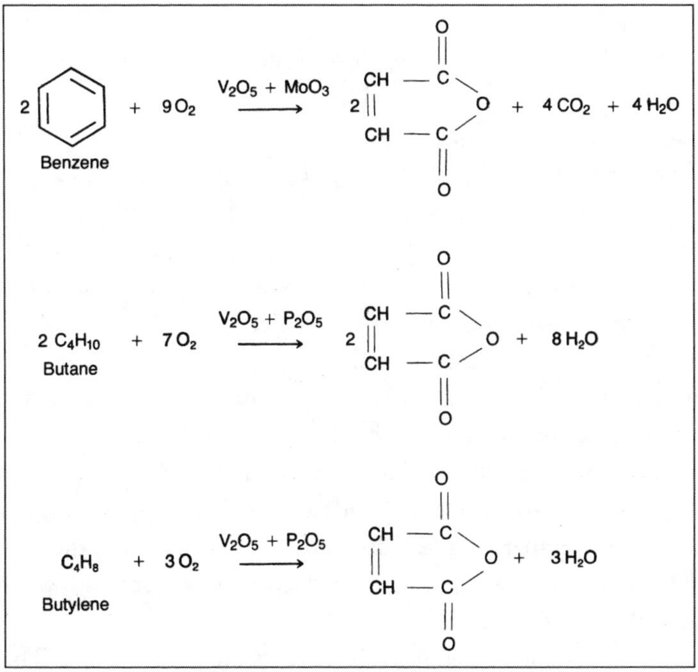

Fig. 20–2 Routes to maleic anhydride

Process

The key to the reactions in Figure 20–2 is the incredible ability the catalyst has to rearrange the atoms and their bonds. The catalyst for all three feedstocks is V_2O_5 (vanadium pentoxide) and a promoter. In the case of benzene, a MoO_3 (molybdenum trioxide) promoter is added. For butane and butylene, it's P_2O_5 (phosphorous pentoxide.) Consider the changes that take place with benzene in only a single pass through the oxidation reactor, a lapsed time of one second:

- The benzene ring is opened, and two of the carbon atoms are cleaved off as CO_2
- The resulting butene molecule undergoes selective oxidation where the two terminal methyl groups are converted to carboxylic acid groups
- A molecule of water is then lost, giving rise to the heterocyclic anhydride grouping and thus, MA

All of this occurs without oxidizing the reactive double bond.

If you set out to accomplish all that in a chemical process, you surely wouldn't expect to be lucky enough to find something as selective and powerful as V_2O_5. Clarke's Third Law is true: "Any sufficiently advanced technology is nearly indistinguishable from magic."

Fixed bed plants. In this type of plant, the process flow for all three feeds looks like the plant in Figure 20–3. The feed and compressed air are mixed, vaporized in a heater, and then charged to the fixed-bed reactor, a bundle of tubes packed with the catalyst. The ratio of air to hydrocarbon is generally about 75:1 to keep the mixture outside the explosive range, always a good idea. The feed temperature is 800–900°F, depending on the feed. The reaction time is extremely quick, so the feed is in contact with the catalyst for only 0.1 to 1.0 second.

Like most oxidation reactions, this one is strongly exothermic. That's why the catalyst is in tubes—coolant is pumped past the tubes to keep the reaction temperature from running away. Also, the reaction is self-sustaining. That is, the reaction gives off enough heat to keep itself going. Once it gets hot enough to get started, it'll continue with no more heat added. The reaction temperature is maintained at 700–900°F.

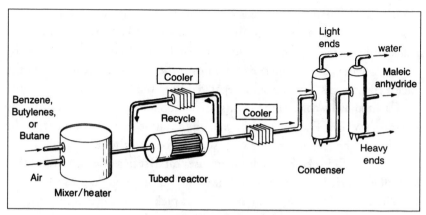

Fig. 20–3 Maleic anhydride plant

The effluent gas from the reactor contains about 50% maleic acid (not maleic anhydride). The balance is some unreacted feed, CO_2, water, and some miscellaneous waste products. A recycle stream is passed through a cooler and recharged to the reactor. The purpose is not only to take another pass at the feed but also to dilute the feed with some already-made maleic acid. That helps to disperse the heat of reaction and to control the operating conditions.

The bulk of the effluent is run through a cooler (heat exchanger) and a condenser to remove the light ends that include traces of carbon monoxide and carbon dioxide and by-product water. The bottom stream is maleic acid, which is easily dehydrated, as in Figure 20–4, by vacuum distillation or azeotropic distillation with ortho-xylene. (*See* Chapter 3 if you've forgotten totally everything about azeotropic distillation.) The dehydrated maleic acid is maleic anhydride. Further purification is done by distillation.

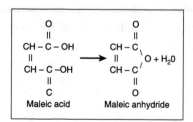

Fig. 20–4 Dehydration of maleic acid

Instead of the two-step, feedstock-to-acid-to-anhydride, another process uses an organic solvent for continuous anhydrous (waterless) recovery of maleic anhydride coming from the reactor. In the recovery section of the process, a patented organic solvent removes the maleic anhydride from the reactor effluent gas. The solvent is separated from the MA later by distillation.

Fluidized bed plants. Some plants use a technology that differs from the scheme shown in Figure 20–3 in that the catalyst moves around with the feed during the reaction rather than staying fixed in the reactor tubes. The design, called fluidized bed technology, uses catalyst in a powdered form that is so mobile that it can be pumped like a liquid or blown like a vapor.

The catalyst is mixed with air where it picks up oxygen atoms, then is blown together with butane into a reactor where the chemical reaction takes place. The effluent from the reactor is a mixture of catalyst, MA, water vapor, and a little feedstock. The catalyst is removed by a contraption called a cyclone, which uses centrifugal force to spin the solid, heavier catalyst particles out of the mixture. The MA and feedstock (butane, etc.) are then separated for recovery (the MA) or recycle (the feedstock).

Commercial aspects

Uses. About 60% of the MA produced is used to make unsaturated polyester and alkyd resins, which are formed by reaction of MA with glycols. Polyester resins are used in the fabrication of glass fiber reinforced parts. Applications include boat hulls, automobile body parts, patio furniture, shower stalls, and pipe. Alkyd resins are mostly used in coatings (paint, varnish, lacquers, and enamels). MA also is widely used as a chemical intermediate in the manufacture of plasticizers and dibasic acids (fumaric, maleic, and succinic). About 15% of MA production goes into the manufacture of viscosity index improvers and dispersants used as additives in lube oils. Several agricultural chemicals are based on maleic anhydride, the best known being Malathion.

Maleic Anhydride Properties

Molecular weight	98.06
Freezing point	127.4°F (53.0°C)
Boiling point	391.5°F (199.7°C)
Specific gravity	0.934
Physical appearance	white needles or flakes with an acrid odor.

Properties, grades, and handling. MA melts at 127°F, so at normal temperatures it is a white solid with an acrid odor. The vapors are highly toxic and will burn your eyes and give you a skin rash. It's soluble in water and many organic solvents.

MA is available commercially in 99% purity in both molten (liquid, above 127°F) and solid forms (flakes, pellets, rod, or briquets). MA is often shipped in fiber drums or bags. Heated tank cars or trucks are used for liquid shipments. Because of the toxic fumes, the hazardous materials designation must be posted on all shipments.

Chapter 20 in a nutshell ...

Maleic anhydride is a cyclic anhydride with one double bond in the ring and two double-bonded oxygens hanging off the ring. The resulting reactivity leads to maleic's use in making polymers, unsaturated polyesters, alkyd resins, plasticizers, and dicarboxylic acids.

Maleic can be made by oxidation of butane or benzene. The process would otherwise be virtually impossible without the use of vanadium pentoxide as the catalyst. It enables extensive reconfiguration of either feedstocks molecular structure into the anhydride structure.

EXERCISES

1. _____, _____, and _____ are three important anhydrides derived from what feedstocks? What are their structural differences, and why is one of these anhydrides more reactive?

2. Clark's Third Law could apply to the highly selective V_2O_5 catalyst for
   manufacturing MA. Name two other catalysts that might fall under
   Clark's Third Law.

3. What are the two good reasons for choosing between butylene, butane,
   or benzene as the feedstock for MA?

CHAPTER 21 ⬡

Alpha Olefins

"I do not view the process with any misgivings."

Winston Churchill,
Tribute to The Royal Airforce
House of Commons
August 20, 1940

Alpha olefins occupy a strangely unique niche in the petrochemicals industry. Their name and their chemical structure imply they are a basic building block. In a way they are. But they are also derivatives of ethylene, and they are grown almost in a way that polymers are grown, just not as long. One other fact that might be surprising is that one of the alpha olefins is butene-1, a petrochemical covered in Chapter 6, The C_4 Hydrocarbon Family. The reason the name *alpha olefin* didn't come up is that, generally, those C_4's result from cracking larger molecules. Although the alpha olefin, butene-1, and all the other alpha olefins once came from this route, now they come from just the inverse type of process—they're grown from the bottoms up.

Alpha olefins are straight-chain hydrocarbons having a double bond in the number one carbon-carbon position. That's called the alpha position, and hence the name alpha olefin. (There are beta, gamma, etc., compounds around, too.) The chains can have as few as four carbons (butene-1) or more than 30 (written C_{30}^+). And they all have the double bond in the alpha position, as you can see in Figure 21–1. At one time, alpha olefins could have either an odd or even number of carbon atoms, but through a quirk of the manufacturing processes now in vogue, only even number carbon count alpha olefins now predominate.

The interest in alpha olefins as a group lies in the reactivity of the double bond, just like styrene, vinyl chloride, ethylene, propylene, or acrylonitrile. But individually, the alpha olefins of varying chain length have quite different physical characteristics and therefore different applications. For example, the C_4 alpha olefin is a gas at room temperature, while the C_6 through C_{18}'s are liquids and the C_{20}^+ are waxy solids.

$CH_3-CH_2-CH=CH_2$ $CH_3-CH_2-CH_2-CH_2-CH=CH_2$

Butene-1   Hexene-1

$CH_3-CH_2-(CH_2-CH_2)_n-CH=CH_2$ where n = 2, 3, . . . to 13+

C_8 to C_{30}+ alpha olefins

Fig. 21–1 Alpha olefins

HISTORICAL DEVELOPMENT

In the early 1960s, alpha olefins were produced by thermally cracking waxy paraffins found in crude oils. The process consisted in subjecting the wax to high enough temperatures to cause cleavage of the carbon-to-carbon bonds in the long wax chain molecules. Because of the absence of extra hydrogen, the cracking process leaves a double bond at the end of the resulting molecules. The various chain lengths were then separated by distillation. This route worked okay but was ripe for improved efficiency. It was energy intensive so it was expensive. It also resulted in a high proportion of branched olefins (having side-chains) because of the characteristics of the feedstocks. If the feed had side-chains, the alpha olefins were likely to have side-chains.

In the late 1960s, the oligomerization route was introduced. (Oligomer comes from the Latin root *olig-,* meaning a few; and *mer,* meaning part, as in monomer, polymer, etc.) The process is based on "growing" chains by addition of ethylene molecules. That overcame the problem of branching and left the expensive, energy-intensive processing to the olefin producers. Within a few years, the oligomerization process had completely replaced all the older technology.

But growing oligomers with ethylene results in chains of only even numbers of carbons. Wax cracking gave both even and odd numbers. Wouldn't that be an obstacle to full commercialization of the oligomerization process? As any tourist guide in a third world country will tell you, "No problem!" As it turns out, the demand for alpha olefins is almost entirely for ranges of chain lengths, not specific carbon count chains. Examples are shown in Table 21–1. On exception is butene-1, which is used by itself in several applications, such as polybutylene manufacture and as a polyethylene comonomer.

By 2000, the alpha olefin market had grown to more than 3 billion pounds. Technology had brought down the cost of producing them and simultaneously, a broad range of applications for all the alpha olefins expanded rapidly—surfactants, synthetic lubricants, plasticizer alcohols, fatty acids, mercaptans, comonomers, biocides, paper and textile sizing, oil field chemicals, lube oil additives, plastic processing aids, and cosmetics.

C_4-C_8	polymers and polyethylene comonomer
C_6-C_8	low molecular weight fatty acids and mercaptans
C_6-C_{10}	plasticizer alcohols
C_{10}-C_{12}	synthetic lubricants and additives, detergent amine oxides and amines
C_{14}-C_{16}	detergent alcohols and nonionics
C_{16}-C_{18}	lube oil additives and surfactants
C_{20}-C_{30} +	oil field chemicals and wax replacement

Table 21–1 Alpha olefin applications

Manufacturing alpha olefins

Ethylene oligomerization is accomplished by successive addition of ethylene molecules.

$$CH_2=CH_2 + CH_2=CH_2 \longrightarrow CH_3-CH_2-CH=CH_2$$
Ethylene   Ethylene   Butene-1

$$CH_3-CH_2-CH=CH_2 + CH_2=CH_2 \longrightarrow CH_3-(CH_2)_3CH=CH_2$$
Butene-1   Ethylene   Hexene-1

$$CH_3-(CH_2)_3-CH=CH_2 + CH_2=CH_2 \longrightarrow \text{ etc., up to } C_{30}{}^+$$
Hexene-1   Ethylene

The routes to commercial processes for these reactions came in three waves:

- The Ziegler process based on a triethyl aluminum catalyst
- The Shell process based on a nickel phosphine catalyst
- The Alpha Select process based on metallocene catalysis

The last, the Alpha Select process, was introduced to satisfy the demand for plasticizer olefins, which was growing faster than the detergent olefins.

Ziegler process. The chemistry of the Ziegler catalyst route to alpha olefins is the same as you read in Chapter 15, The Higher Alcohols. The treatment here is another approach, and you might find it instructive. Or not.

Karl Ziegler, the notable German chemist discovered the catalyst that ignites the process of linking ethylene molecules in a straight chain. Ziegler found that triethyl aluminum could, under the right pressure and temperature conditions, be used as a kind of a root for growing hydrocarbon chains. Triethyl aluminum is a compound of aluminum with three ethyl groups attached. When subjected to high pressures and temperatures and an excess of ethylene, a hydrocarbon atom at the terminal end of the ethyl group can be displaced by ethylene, starting the growth of a chain. Other ethylene molecules will also continue adding at the end of the new chain, as long as there are sufficient ethylene molecules around and the temperature and pressure conditions are right.

When the process of chain growth is satisfactorily completed, separation of the three hydrocarbon chains that are connected to the aluminum atom is accomplished by a displacement reaction. The chain-laden aluminum compound (called trialkyl aluminum compounds) is subjected to still higher temperatures and pressure. This causes an ethylene molecule to displace the long linear carbon chain. As the separation is made, triethyl aluminum is reformed, making a recyclable root for another go-around.

The displacement reaction:

$$Al[CH_2\text{-}CH_2\text{-}(CH_2\text{-}CH_2)_n\text{-}CH_2\text{-}CH_3]_3 \;+\; CH_2{=}CH_2 \longrightarrow$$

Aluminum alkyl   Ethylene

$$Al(CH_2\text{-}CH_3)_3 \;+\; 3CH_3\text{-}CH_2\text{-}(CH_2\text{-}CH_2)_n\text{-}CH{=}CH_2$$

Triethyl aluminum   an Alpha olefin

This chemistry is sometimes accomplished simultaneously in one reactor and sometimes in two separate reactors. In the former, the triethyl aluminum catalyst is lost; in the latter, it is recycled. Sometimes the displacement compound is butene-1 or hexene-1, depending on the chain length of the final alpha olefin desired and the change in operating conditions necessary to effect the displacement reaction.

Typical yields of the two processes are shown in Table 21–2. Actually there is more flexibility than what the table suggests. In the early 1970s, most of the demand for alpha olefins was in the C_{10}–C_{16} range. That was at a time when environmental concerns were escalating over the foaming being caused by phosphate-based household detergents. As a consequence, the demand for biodegradable detergents that happened to be based on linear, even-numbered C_{10}–C_{16} range alpha olefins precursors took off. The demand skewed toward the middle of the range of alpha olefin production. But by continuous development of the process, the production of alpha olefins at either end of the distribution was mitigated. There is continuing progress in matching production to distribution, even as new applications of both ends of the range evolve.

	One-step process	Two-step process
C-4	13–16	5–10
C-6 to C-10	50–40	42–63
C-12 to C-18	27–30	25–50
C-20 +	10–14	2–3

Table 21–2 Alpha olefin production distribution (percent by weight)

The process

The general flow and the reactions for a typical alpha olefin process are shown in Figure 21–2.

High purity ethylene gas plus recycle ethylene are fed to a compression chamber, compressed and then fed along with catalyst previously dissolved in a suitable solvent into parallel horizontal reactors, as many as eight in parallel. Each reactor consists of a water-filled shell containing a single pipe, coiled to give maximum contact with the water. Reaction conditions are 350–425°F and 2000–3000 psi.

The reaction is exothermic, and the heat transfers to the water, creating steam. A backpressure control regulates how much steam is released from the reactor shell. The faster or slower the reaction, the more or less heat transfers, and the more or less steam gets generated. The controls also let in more or less fresh water, all this controlling the reaction temperature and rate.

Olefin conversion per pass can be as high as 60% with yields approaching 95%. (Conversion relates to how much ethylene disappears in one pass; yield relates to how much of it ends up in the finished product, not in the by-products. See the appendices for further discussion.)

The reactor effluent is cooled and fed to the ethylene separator for recovery of unreacted gaseous ethylene. The liquid phase is filtered to remove small amounts of polymer and then treated with aqueous caustic to remove the catalyst. The dissolved light ends (C_2 and C_4 olefins) are separated by suitable fractionating towers in series. A portion of the ethylene is purged to remove methane and ethane, and the remaining ethylene is recycled to the compressor. The butene-1 is removed to storage.

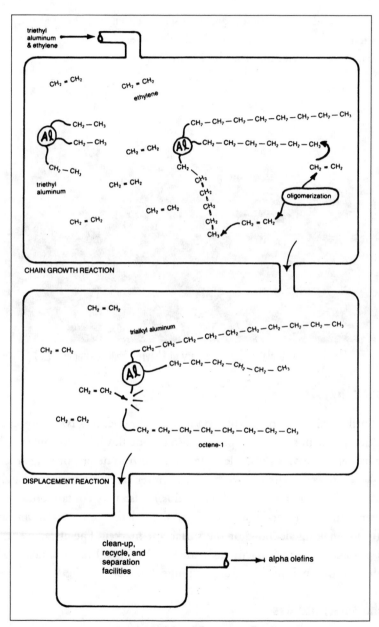

Fig. 21–2 Alpha olefin process flow

The C_6 and heavier olefins are then separated via a series of atmospheric and vacuum fractionation towers. Multiple towers or columns are required to separate the heavier olefins. (*See* Figure 21-3.)

Fig. 21-3 Alpha olefin plant (courtesy of Chevron Phillips Chemical Co.)

Other catalysts

Shell developed a nickel/phosphine catalyst, used in a two-step process. The displacement step in this process also has the flexibility to convert whatever lighter or heavier alpha olefins are created into detergent range olefins, C_{10}–C_{16}. However, the double bonds are internal rather than in the alpha position. (They are not, sniff, alpha olefins.) Also they contain odd- as well as even-number carbon chains. These straight chain olefins are nevertheless useful for making alcohols for the surfactant market. The attractiveness of this process is the manufacturer's choice of having all of the linear olefins (alpha and internal) in the detergent range, C_{10}–C_{16}.

Alpha select process

The dramatic growth of the lower range alpha olefins was greatly accommodated by the development and commercialization of a new approach in the 1990s, the Alpha Select process. (*See* Figure 21-4.) The jump-start in this process comes from metallocene catalysts, organometallic complexes based on, in this case, titanium, zirconium, vanadium, nickel, or palladium metal and alkyl aluminum oxides. The three-step process—initiation, propagation, and termination—is very similar to one of the classic processes for making polymers such as polyethylene and polypropylene.

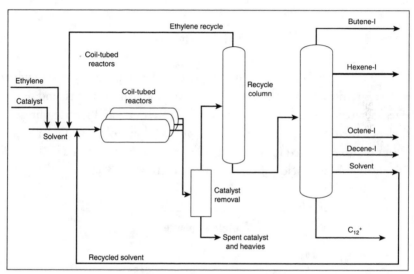

Fig. 21–4 Alpha select process

The process operates in the liquid phase by dissolving the ethylene in an inert solvent such as cyclohexane or isopentane. The metallocene catalyst is also injected to the mix. The solvent has several important functions. It keeps in solution the alpha olefins produced as well as the ethylene and catalyst. It also enhances the catalyst activity and selectivity.

The ethylene- and catalyst-laden solvent is injected continuously into the reactor at 250–300°C and 1300 psi where the ethylene molecules react to create almost exclusively C_4, C_6, C_8, and C_{10}, straight-chain alpha olefins in the proportions shown in Table 21–3.

% by Weight	
Butene – 1	33–43
Hexene – 1	30–32
Octene – 1	17–21
Decene – 1	9–14

Table 21–3 Alpha olefin product distribution

The reactor effluent is fed to the spent catalyst separation section where catalyst is removed, treated to remove any volatile hydrocarbons, and sent to be regenerated. The effluent is then distilled to remove and recycle unreacted ethylene, then fractionated into high purity alpha olefins. The reaction solvent is also recovered for recycling. Olefin conversion per pass is 50–60%, with the combined yields of C_4–C_{10} alpha olefins of 93%.

Material Balance	
Feed:	
Ethylene	2684 lbs.
Product:	
C_4–C_{10} alpha olefins	2200 lbs.
C_{12}^+ olefins	308 lbs.
Heavy ends	176 lbs.

Properties and handling. Typical properties for the alpha olefins produced by ethylene oligomerization are given in Table 21–4. You can find in the table that as the carbon count increases, purity declines. The impurities are branched chains and internal olefins (beta, gamma, etc.) These variations have more opportunity to form as the molecules get longer—Murphy's Third Law in operation again.

	Butene-1 C_4	Hexene-1 C_6	Octene-1 C_8	Decene-1 C_{10}	Do-decene-1 C_{12}	Tetra-decene-1 C_{14}	Hexa-decene-1 C_{16}	Octa-decene-1 C_{18}	C_{20}-C_{24}	C_{24}-C_{28}	C_{36}+
% Purity	99.0	97.5	96.5	94.5	94.5	93.5	92.0	91.0	ca. 88	ca. 86	ca. 84
Distillation Range — °F	43.0	144-147	248-257	334-347	401-428	454-491	518-572	—	—	—	—
°C	6.1	62-64	120-185	168-175	205-220	240-255	270-300	—	—	—	—
Color and Appearance	colorless, clear							white, waxy solid			
Freeze Point °F	—	—	—	—	—	—	40	65	110	145	180
Flash Point °F	—	less than 60		120	180	225	270	290	375	380	510
Specific Gravity	0.595	0.678	0.718	0.745	0.763	0.776	0.785	0.792	0.799	0.819	0.830
Weight per Gallon, lbs.	5.00	5.70	6.04	6.26	6.41	6.52	6.59	6.65	6.71	6.88	6.95

Table 21–4 Typical alpha olefin properties

The flash points shown are standard measures of flammability. They are the temperatures at which the liquid gives off enough flammable vapor to the surrounding air to ignite. The lower the flash point, the more flammable and dangerous the compound. The products are shipped in tank trucks, tank cars, and barges that may or may not have heating coils. The higher carbon number alpha olefins, C_{18}^+, will require heating coils. Nitrogen is required for the liquid C_6 to C_{16}s because exposure to air gradually causes a reaction with oxygen, producing peroxides. Stabilizers also are used to minimize peroxide buildup.

DOT classifies butene-1 as a flammable compressed gas, hexene, and octene as flammable liquids, and decene-1 as a combustible liquid. All four must carry required shipping placards. The remaining alpha olefins are not regulated by the DOT.

Commercial Aspects

The lion's share, 80–90%, of the alpha olefins produced in the United States are used in five areas: as comonomers in LLDPE (linear low density polyethylene) and HDPE (high density polyethylene), plasticizer alcohols, polyalpha olefins for use in synthetic lubricants, detergent alcohols, and surfactants. The comonomer demand started out exclusively as butene-1, but it is shifting toward hexene-1 and octene-1. Similarly, the specification for decene-1 in polyalpha olefins is relaxing to include octene-1 and dodecene-1. The remaining 10–20% of production is used in the other applications listed in Table 21–1.

Chapter 21 in a nutshell ...

Alpha olefins are straight-chain olefins that have a double bond in the number one (alpha) carbon-carbon position. Because they are now made by linking ethylene molecules together, alpha olefins have only even-number carbon counts. Alpha olefins with 4, 6, 8, to 30 or more are commercially available.

Alpha olefins are made either by oligomerization, growing them on an aluminum root by adding ethylene until the desire size is reached, or by catalytic processes, one favoring the shorter alpha olefins.

The variety of alpha olefin application is extensive, including polymers, surfactants, synthetic lubricants, lube oil additives, plasticizer alcohols, mercaptans, and fatty acids.

EXERCISES

1. What is the difference between the Ziegler process for producing alpha olefins and the one described in Chapter 15 for producing higher alcohols?

2. Why are there only even-number carbon count alpha olefins any more?

3. In the Ziegler process, why do you think there is a distribution of different carbon count alpha olefins rather than just one?

4. What is the main difference in results between the Ziegler process and the Alpha Select process?

COMMENTARY 3 ⬡

"It is the beginning
of the end."

Charles Maurice de Talleyrand, 1754–1831
(Commenting on the Battle of Bordino in 1812)

REVIEW

There are many areas of commonality in the hodgepodge of petro-chemicals just covered. Synthesis gas, the underground petrochemical building block, is the route to some alcohols including methyl, isobutyl, and normal butyl alcohol. Yet hardly anyone knows anything about synthesis gas because it's not a traded commodity. But you can't get much more basic a building block than carbon monoxide and hydrogen.

Synthesis gas is also the precursor to MTBE via methanol. The process requires isobutylene as well. Ethyl alcohol is made by direct, catalyzed hydration of ethylene. The route to isopropyl alcohol historically used to be sole-ly indirect hydration of propylene, which occurs at much lower pressures and temperatures than the direct method, but advances in catalysis now make the direct route competitive.

The process to make 2-ethyl hexanol starts with propylene and synthe-sis gas, but it takes dimerization and hydrogenation to form the proper car-bon chain and alcohol group, -OH.

The processes for the ketones, acids, acrylonitrile, and the acrylates, and maleic anhydride defy simple summarization. Just read them individually. The chemical structures for most of them are shown in the following table. It might help if you try to recognize the signature group in each molecule.

As one of the newest petrochemicals to make the big time, alpha olefins have found two niches. One is in the petrochemicals industry as precursors for detergents, copolymers, and specialty chemicals. The other niche is in this book, appropriately sandwiched in between the usual list of petrochemical derivatives and the polymers. The alpha olefins are indeed derivatives, but the process for creating them is more like polymerization than any other derivative process.

FOREWORD

The best is yet to come. The most interesting chapters in this book are the next few. One reason is that the polymers are so much a part of your everyday life. Much of what you touch and see nowadays (and even what you are) is made in whole or in part from polymers. So in the next few chapters you'll be coming across words that you were familiar with before you ever got into the petrochemical business.

The other reason is that polymer chemistry pulls together a lot of what you've already learned about petrochemicals. So many of them end up in the polymerization processes.

The polymer chapters tend to be long. There's a lot to cover under each topic. As a matter of fact, before you get to read about the polymers in Chapters 23 and 24, you need to read about the nature of polymers in Chapter 22. It's a big body of chemistry and chemical engineering, but these chapters should give you a handle on it.

Chemical Structures of Various Compounds

Name	Nickname	Chemical configuration
Synthesis gas	Syngas	H-H and CO

Alcohols

Name	Nickname	Chemical configuration
Methyl alcohol	Wood alcohol	CH_3-OH
Ethyl alcohol	EA	CH_3-CH_2-OH
Isopropyl alcohol	IPA	$CH_3-CH-CH_3$ $\overset{|}{OH}$
Normal butyl alcohol	NBA	$CH_3-CH_2-CH_2-CH_2-OH$
2-ethyl hexanol	2-EH	$CH_3-CH_2-CH_2-CH_2-CH-CH_2-OH$ $\overset{|}{CH_2-CH_3}$

Ketones

Name	Nickname	Chemical configuration
Acetone	DMK	$CH_3-\overset{\|}{\underset{O}{C}}-CH_3$
Methyl ethyl ketone	MEK	$CH_3-\overset{\|}{\underset{O}{C}}-CH_2-CH_3$
Methyl isobutyl ketone	MIBK	$CH_3-\overset{\|}{\underset{O}{C}}-CH_2-\overset{}{\underset{CH_3}{CH}}-CH_3$

Acids

Name	Nickname	Chemical configuration
Acetic acid	—	$CH_3-\overset{\|}{\underset{O}{C}}-OH$
Acrylic acid	—	$CH_2 = CH-\overset{\|}{\underset{O}{C}}-OH$
Adipic acid	—	$HO-\overset{\|}{\underset{O}{C}}-(CH_2)_4-\overset{\|}{\underset{O}{C}}-OH$
Terephthalic acid	TPA, PTA	$HO-\overset{\|}{\underset{O}{C}}-\bigcirc-\overset{\|}{\underset{O}{C}}-OH$

Chemical Structures of Various Compounds (cont.)

Name	Nickname	Chemical configuration

Acrylonitrile Family

Name	Nickname	Chemical configuration
Acrylonitrile	Acrylo	$CH_2 = CH\text{-}CN$
Methyl acrylate	—	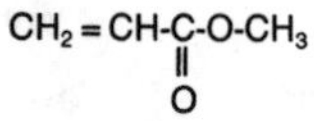 $CH_2 = CH\text{-}\overset{\displaystyle O}{\underset{\|}{C}}\text{-}O\text{-}CH_3$

Methyl methacrylate, MMA:

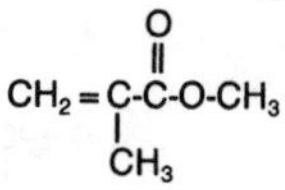

$$CH_2 = \underset{\underset{\displaystyle CH_3}{\|}}{\overset{\overset{\displaystyle O}{\|}}{C}}\text{-}C\text{-}O\text{-}CH_3$$

Anhydrides

Phthalic anhydride —

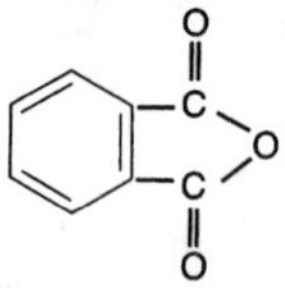

Maleic anhydride, Maleic:

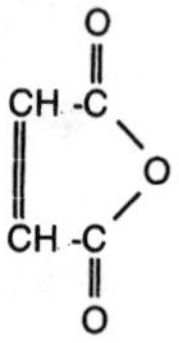

Alpha Olefins

Name	Nickname	Chemical configuration
Hexene-1	—	$CH_2 = CH\text{-}CH_2\text{-}CH_2\text{-}CH_2\text{-}CH_3$

CHAPTER 22 ⬡

The Nature of Polymers

"These are ties which,
though light as air,
are links of iron."

On Conciliation With America
Edmund Burke, 1729–1979

Polymers are a pretty complicated subject. That's why they're treated in three successive chapters. In this one you'll find a number of ways people classify polymers. It's quite an inventory:

Resins vs. Plastics
Thermoplastics vs. Thermoset
Homopolymers vs. Copolymers
Bifunctional vs. Polyfunctional
Linear vs. Branched vs. Cross-Linked
Addition vs. Condensation

The problem is that polymer chemistry became a virtual explosion of ideas and options as it developed in the 1950s and was further commercialized in the 1960s. There's no easy way to cover polymers other than to wade through. But go ahead. It's not hard, and you'll learn a lot.

A little history

The first partially synthetic polymer dates back to 1869, when cellulose (wood pulp) was nitrated (nitrocellulose). The cellulose became processible, and with the further addition of camphor (which acted as a plasticizer), it became a clear, tough, moldable product with the trade name "Celluloid." It was widely used at the end of the 19th century in the form of combs, brushes, photographic film, and shirt collars.

Not much commercial development took place until the chemistry of polymerization was starting to be understood in the 1930s and 1940s. Commercialization of some of the key polymers happened as follows:

1869—Nitrocellulose
1908—Bakelite (first synthetic commercial plastic)
1919—Polyvinyl acetate
1931—The polyacrylates
1936—Polyvinyl chloride
1938—Nylon and polystyrene
1942—Polyethylene and polyesters
1947—Epoxies
1953—Polyurethanes
1957—Polypropylene
1964—Polyimides
1973—Polybutylenes
1977—Linear low density polyethylene

More than 50% of the chemical industry in the United States is now based on or dependent on polymers.

Classifying polymers

For a field of scientific and engineering endeavor, polymers have one of the more sloppy sets of nomenclature. Ask six people in the business to give you definitions of resins and plastics, and you'll get at least six different answers. Almost everyone will tell you that they're both polymers, and that's right. Some will tell you they're interchangeable. Strictly speaking, they're wrong.

A lot of people will tell you that plastics will flow when they're heated or reheated, but that resins are set permanently so that heating them won't do anything. A lot more people, particularly in the fabrication end of the business, think resins are unfabricated polymers, and plastics are resins after they've been molded and/or set by the process, extruders, etc.

If you trace the word resin back far enough, you'll find that it was originally defined as a low molecular weight, natural polymer that is an exudate of (it exudes from) vegetable or non-vegetable matter. Examples are rosin (from pine trees), shellac (from insects), and both frankincense and myrrh (aromatic gums from an East African and an Asian species of tree). Resins like these do not flow if heat and pressure are applied, like plastics do. They decompose or melt. (This definition of resin is obsolete in commerce today.)

So you'll get no neat definition of resins and plastics here. But you'll know to be careful when someone else uses either term. Now, as for polymer, that's defined as a high molecular weight molecule formed by joining, in a repetitive pattern, one or more types of smaller molecules.

Polymers fall into one of two major classes, thermoplastics, and thermosets. Despite the fact that thermosets have been around much longer, thermoplastics make up about 80% of the industry output. Thermoplastics are linear polymers that can be resoftened a number of times, usually by applying heat and pressure. They can be dissolved in solvents (suitable for that purpose). That's not true for thermosets once they're set. After they're formed or cured (by heat and/or pressure), these cross-linked three-dimensional polymers become nonmelting and insoluble. Thermosets actually decompose under heat before they melt.

Both thermoplastics and thermosets can be used in four of the five major application areas: plastics, elastomers, coatings, and adhesives. But, only thermoplastics can be used in making fibers. During the spinning and drawing process of fiber processing, it's necessary to orient the molecules. Only unbranched, linear polymers (not thermosets) are capable of orientation.

Polymers result from polymerization—the chemical combination of a large number of molecules of a certain type, called monomers. Monomers can be bifunctional (capable of joining up with two other monomers) and tri- or polyfunctional (each may join up with three or more monomers). When bifunctional monomers react with each other, you get linear thermoplastic polymers. If tri- or polyfunctional monomers react, you get cross-linked polymers, most of which are thermosets. Figure 22–1 illustrates these variations.

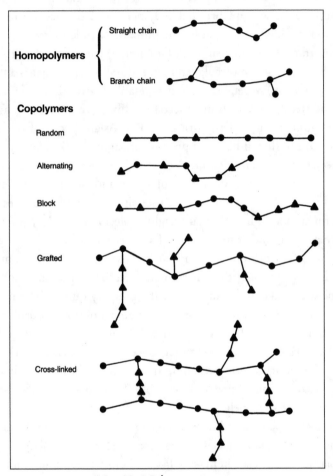

Fig. 22–1 Polymer structures

In some cases, the monomers react with themselves to form homopolymers:

- ethylene to polyethylene
- vinyl chloride to polyvinyl chloride
- styrene to polystyrene

In other cases, and actually most of the time, two or more different monomers react to form copolymers:

- butadiene and styrene to Buna S Rubber
- styrene and acrylonitrile to SAN
- ethylene glycol and terephthalic acid to polyethylene terephthalate (PET)

Making polymers

The polymerization process can be an addition reaction or a condensation reaction. Addition involves monomers containing a carbon-carbon double bone like this: $CH_2=CH-R$. If R is hydrogen, then the monomer is ethylene. If R is chlorine, then the monomer is vinyl chloride; if it's a methyl group, then the monomer is propylene; a benzene ring, then it's styrene; and so on to more complicated structures.

Condensation polymers generally result from simple reactions involving two different monomers, each containing different functional groups. The usual example is terephthalic acid and ethylene glycol to make polyester. The two monomers will react in such a way that a small molecule like water or methanol is given off a by-product

Addition polymerization

This type of polymerization is a technique for adding monomers end to end. It involves three steps: initiation, propagation, and termination.

Initiation. The trick here is to get the reaction started. Usually a catalyst is used, typically an organic peroxide such as ditertiary butyl hydroperoxide. Peroxide molecules are somewhat unstable, and when they're heated, they decompose and turn into highly reactive free radicals. As you'll recall, a radical is an almost-complete molecule, but all the valence requirements are not satisfied. So it is very anxious to meet up with some other molecule to satisfy its valence needs. The free radical, in the presence of an abundance of monomers, say a million to one ratio, will react with a monomer molecule. It becomes part of the molecule. In doing so, the unsatisfied valence condition now transfers to the end of the monomer. A new radical is formed. That's the start of the initiation step.

Propagation. Now the new radical collides and reacts with another monomer to give a new larger radical, which in turn reacts with another monomer, and so on, and so on. This chain growth continues until propagation is terminated. The propagation or growth step in a commercial process usually takes a couple of seconds. The number of monomers in the chain is at least a 1000 or more. By employing special catalysts, some polyethylenes are produced with up to 150,000 repeating units, and all in a few seconds.

Termination. A number of mechanisms are used to stop the propagation or growth step. A common way occurs when the monomer concentration is so low that the free radical chains dimerize. That is, they collide with each other and form a stable polymer, with all valence requirements satisfied.

Branch polymers or branch chains are short or long chains which are at right angles to the original chain's backbone. (*See* Figure 22–1 again.) Short chains can be deliberately added using comonomers such as butene-1 or hexene-1. Long-chain branching often happens in high-pressure polyethylene processes. In the propagation step, a "growing" polymer radical extracts an "inside" hydrogen atom from a "finished" polymer chain. That now becomes a new polymer radical (at that site), and a chain can start growing there. Sometimes this new reaction is facilitated by a chain transfer agent. Isobutane, propylene, and dodecyl mercaptan do well.

Cross-linked polymers occur when polymer chains are linked together at one or more points (other than their ends). Cross-linking can occur when the monomers involved are polyfunctional. That is, they have more than two active sites where links can be attached. So they grow like long chains, but they also link up with each other. Cross-linking can also be initiated by adding special agents. (Like Charles Goodyear did when he accidentally spilled some sulfur into a vessel of molten natural rubber. In the process, he "discovered" vulcanization, cross-linking with sulfur atoms.) Cross-linked polymers lose their moldability, even when they're reheated because the molecules are chemically bonded in place and do not slip and slide.

The length, branching, and cross-linking of the polymers are controlled by the timing of the three steps. A lot of initiating catalyst will result in an abundance of free radicals. When that happens, the concentration of the monomer goes down rapidly as a relatively high number of polymers start growing all at once. This results in early termination and a large number of small (low molecular weight) polymers. The properties of these polymers would be very different (maybe better or worse) than the converse, a relatively small number of large (high molecular weight) polymers. Usually, the large molecules are what you're after.

Copolymers. Mixtures of two or more different bifunctional monomers can undergo additional polymerization to form copolymers. Why copolymerize? Well, polymers have different properties that depend on their composition, molecular weight, branching, crystallinity, etc. Many copolymers have been developed to combine the best features of each monomer. For example, polystyrene is low cost and clear, but it is also brittle with no toughness. It needs internal plasticization. By copolymerizing styrene with a small amount of acrylonitrile or butadiene, the impact and toughness properties are dramatically improved.

Another reason for copolymerization is to insert functional grouping in the polymer. A functional group is one that is easily reacted. For example, copolymerization of styrene with acrylonitrile, $CH_2=CH-CN$, involves only the double bond, leaving the newly formed copolymer with the active functional group -CN, available for subsequent reaction. The copolymer might be reacted later with itself or another monomer to give a cross-linked thermoset.

A third reason for interest in copolymers is crystallinity. Transparency and translucency are greatly affected by crystalline properties, which can be regulated by copolymerization.

Condensation polymerization

Condensation polymers are always copolymers. They are always formed by a series of chemical reactions involving two reactive sites in each that can join to form linking bonds. By-products are usually given off and are generally small molecules such as water, methanol, or hydrogen chloride.

Because two reactive sites are necessary, bifunctional monomers are often used in condensation polymerization. A bifunctional monomer includes molecules with two identical signature groups in them. Examples would be terephthalic acid and ethylene glycol, both shown in Figure 22–2. When a bifunctional monomer like either of those is used, the polymerization step is end-to-end, forming long chains. The reaction in Figure 22–2 is a "simple" esterification of ethylene glycol with terephthalic acid to make polyethylene terephthalate, which is polyester fiber or Dacron.

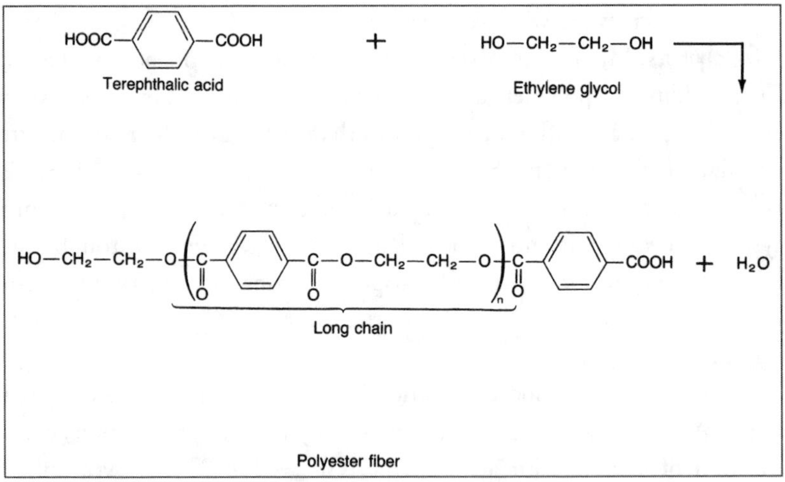

HOOC—⟨benzene ring⟩—COOH   +   HO—CH_2—CH_2—OH

Terephthalic acid   Ethylene glycol

HO—CH_2—CH_2—O$\left[\text{C}-⟨\text{ring}⟩-\text{C}-\text{O}-CH_2-CH_2-\text{O}\right]_n$C—⟨ring⟩—COOH   + H_2O

Long chain

Polyester fiber

Fig. 22–2 A condensation polymerization

There are numerous bifunctional monomers used in condensation polymerization. Some of the more popular signature groups that turn up frequently are shown in Figure 22–3. Important copolymers made by condensation include epoxies, nylon, polyesters, polycarbonate, and polyimides. As always, there are exceptions, and one is Nylon 6 made by a ring opening reaction of caprolactam. All of these will be covered in the next two chapters.

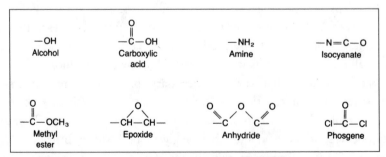

Fig. 22–3 Common signature groups used in condensation polymers

Thermosets

Thermosets cure into nonmelting, insoluble polymers. Frequently, the curing needs heat, pressure, or catalyst to proceed. Often the final cure, which is nothing more than completion of the cross-linking, takes place in the fabrication or molding operations. The chemistry is about the same as you saw in the thermoplastics, but there are more reactive sites per monomer. (They are polyfunctional.) Consequently, more three-dimensional cross-linking takes place.

The simplest way to achieve three-dimensional cross-linking is to use monomers with three or more reactive sites. Examples are maleic anhydride, butadiene, isoprene, epichlorohydrin, pyromellitic dianhydride, and trimethylol propane.

As an illustration, consider some of the elastomers. In its natural state, rubber lacks toughness. In the 1939 accident already mentioned, Goodyear found that by reacting latex rubber (natural or synthetic) with sulfur, he could improve its strength and toughness, and increase its temperature properties. What he was doing was cross-linking rubber with sulfur in a process now commonly called vulcanization. The reaction with polyisoprene rubber is shown in Figure 22–4. Other synthetic rubbers such as butyl rubber (from isobutylene and butadiene), Buna S (butadiene and styrene), Buna N (butadiene and acrylonitrile), and neoprene (chloroprene) can all be vulcanized to thermoset rubbers. They all have the polyfunctional configuration that makes cross-linking by sulfur possible. After vulcanization, they are tough, resist deforming, and are heat- and cold-insensitive, resistant to solvents, and nonconductive.

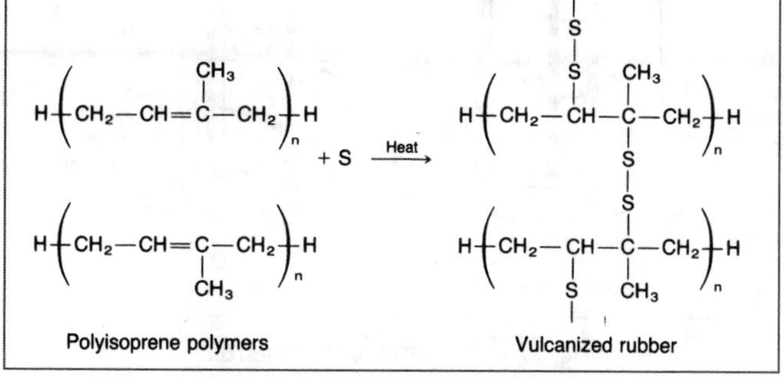

Fig. 22–4 Vulcanization of isoprene rubber

Other important thermosets include phenolics like Bakelite, epoxy resins, polyimides, and polyurethanes.

Methods of polymerization

Most polymers are made by one of four processes commercially available. Each process has advantages related to the monomer being used and the end use of the polymer.

Bulk polymerization. This is the simplest method. Monomers and initiator are mixed in the reactor shown in Figure 22–5 and heated to the right temperature. The bulk process is suitable for condensation polymers because the heat of reaction is low (it gives off less heat).

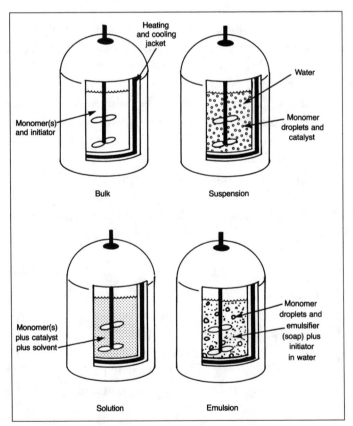

Fig. 22–5 Processes for polymerizing

Polymerization of methyl methacrylate to Plexiglas is done in the bulk process. High pressure polymerization of ethylene is done this way also. But other addition polymerizations frequently become too exothermic and without adequate heat removal system, the reaction tends to "run away" from optimum conditions.

Solution polymerization. Highly exothermic reactions can be handled by this process. The reaction is carried out in an excess of solvent that absorbs and disperses the heat of reaction. The excess solvent also prevents the formation of slush or sludge, which sometimes happens in the bulk process when the polymer volume overtakes the monomer. The solution process is particularly useful when the polymer is to be used in the solvent, say like a coating. Some of the snags with this process: it's difficult to remove residual traces of solvent, if that's necessary; the same is true of catalyst if any is used. This process is used in one version of a low-pressure process for high-density polyethylene and for polypropylene.

Suspension polymerization. In this process, monomers and initiator are suspended as droplets in water or a similar medium. The droplets are maintained in suspension by agitation (active mixing). Sometimes a water-soluble polymer like methylcellulose or a finely divided clay is added to help stabilize or maintain the droplets. After formation, the polymer is separated and dried. This route is used commercially for vinyl-type polymers such as polyvinyl chloride and polystyrene.

Emulsion polymerization. Soap is usually the emulsifying agent. The most useful characteristic of soap is the way the soap molecules behave in contact with oil and water. One end of a soap molecule is oleophilic (oil-loving) and the other is hydrophilic (water-loving). In an oil/water solution, the soap molecules form micelles, tiny structural units, suspended in the water. The oleophilic ends are pointed outward, interacting with the water medium. The polymerization actually takes place within the micelle, which remains suspended in the water. So, as it grows, the polymer remains suspended in the water. Very high molecular weight polymers are produced by this technique in the form of latex. The process is particularly suitable for polymers used in paints, like polyvinyl acetate.

Polymer properties

The proof of the polymer is in its properties. It is the physical properties that the engineers use in selecting polymers. They include density, tensile strength, impact strength, toughness, melt index, creep (ability to elongate), modulus of elasticity, electrical characteristics, thermal conductivity, appearance, flammability, and chemical resistance. Add price and fabricating costs, and the engineer has most of the data he/she needs to select the correct polymer for his application.

Generally speaking, the physical properties of polymers depend on crystallinity, molecular weight, molecular weight distribution, linearity/cross-linking, and chemical composition/structure.

Crystallinity. Is one of the key factors influencing properties. You can think of crystallinity in terms of how well a polymer fits in an imaginary pipe, as in Figure 22–6. Linear, straight chains are highly crystalline and fit very well. Bulky groups, coiled chains, and branched chains are not able to "line up" to fit in the pipe. They are amorphous, the opposite of crystalline. In a spectrum from totally amorphous, to almost totally crystalline, there is methyl methacrylate, polypropylene, low-density polyethylene, linear low-density polyethylene, high-density polyethylene, and nylon.

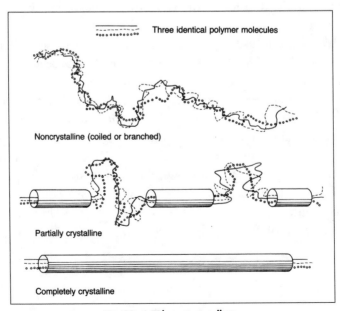

Fig. 22–6 Polymer crystallinity

With increasing crystallinity, polymers tend to be denser. They're not too different from pasta. A pound of uncooked spaghetti fits in a smaller box than a pound of uncooked macaroni. The spaghetti is like a perfectly crystalline structure. The macaroni is the opposite. Along with increasing density comes greater tensile strength, higher softening point, and more opaqueness. Further, both elongation (or stretch) and impact strength decrease with increasing polymer density. The most surprising of these relationships is that the greater the crystallinity, the less translucent it becomes. It doesn't help to ask you if that's "crystal clear," because that expression implies just the opposite.

Molecular weight. The molecular weight influences the melt viscosity, tensile strength, the low temperature brittleness, and the resistance to tearing.

Structure and chemical affect a number of the properties of polymers. These are discussed below.

Thermal stability. The presence of side chains, cross-linking, and benzene rings in the polymer's "backbone" increase the melting temperatures. For example, a spectrum of polymers with increasing melting temperatures would be polyethylene, polypropylene, polystyrene, nylon, and polyimide.

Stress-strain characteristics. Linear chain polymers are quite flexible and subject to creep or stretch. Branching or rings in the "backbone" have a stiffening effect. For example:

Polyethylene	—soft, tough, high creep
Polypropylene	—hard, tough, medium creep
Polystyrene	—hard, brittle, low creep
Cross-linked thermosets	—hard, brittle, no creep

Density. Once something more than C-H is introduced to polymers, most of them get denser. In order of increasing density are polypropylene, polyethylene, polystyrene, polyvinyl chloride, and Teflon.

Flammability. Presence of chlorine, fluorine, bromine, or phosphorous in a polymer reduces flammability. Thermosets are more flame-resistant than thermoplastics.

Moisture absorption. Directly related to the atoms making up the polymer. The more moisture-absorbing the molecule, the less dimensional stability; strength, stiffness, electrical properties are also adversely affected.

Those are the generalities of polymers. The specifics of low- and high-density polyethylene, polypropylene, polyvinyl chloride, and polystyrene are covered in the next chapter and resins and fibers in the last.

Chapter 22 in a nutshell ...

Polymers are high-molecular weight compounds made by joining together hundreds or thousands of molecules. These molecules usually consist of one or two types of petrochemicals called monomers.

Polymers are generally one of two types:

- **Thermoplastics:** These can be dissolved or softened and remolded several times after their initial productions
- **Thermosets:** These set permanently and cannot be remolded and are not meltable or soluble

Thermoplastics are generally long-chain, linear, two-dimensional molecules, while thermosets are generally three-dimensional long chains, connected by cross-linking chemical bonds.

There are two different types of chemical reactions used to make polymers:

- **the addition reaction,** where monomers are added end to end like, for example, polyethylene or polystyrene.
- **the condensation reaction,** where two or more different kinds of monomers are used to form copolymers. They are always formed by chemical reactions involving two reactive sites one on each monomer. Some by-product like water, hydrogen chloride is always formed. An example is ethylene glycol and terephthalic acid to make polyethylene terephthalate (polyester fiber) and by-product water.

The properties of polymers vary considerably, making the match between polymer and application a sort through such characteristics as density, tensile and impact strength, toughness, melt index, creep, elasticity, heat and chemical stability, electrical properties, flammability, and price.

EXERCISES

1. What's the difference between a thermoplastic and a thermoset?

2. What are the four basic steps of addition polymerization?

3. Name four processes generally used for polymerization reactions.

4. Only _____ can be used in making fibers.

5. Which of the following polymer structures are homopolymers, and which are copolymers? What monomers are involved?

   a. $-CH_2-O-CH_2-O-CH_2-O-CH_2-$

   b. $-O-CH_2-CH_2-O-\overset{\overset{\displaystyle O}{\|}}{C}-CH_2-CH_2-CH_2-CH_2-\overset{\overset{\displaystyle O}{\|}}{C}-$

   c. $-\overset{\overset{\displaystyle Cl}{|}}{C}H-CH_2-\overset{\overset{\displaystyle Cl}{|}}{C}H-CH_2-\overset{\overset{\displaystyle Cl}{|}}{C}H-CH_2-$

CHAPTER 23 ⬡

Thermoplastics

"Let there be spaces in your togetherness."

The Prophet
Kahil Gibran, 1883–1931

L ike it or not, it's a plastic world out there. Plastics have penetrated the traditional markets for paper, cotton, wool, wood, leather, glass, metals, and concrete. (It's a good thing you can't eat it.) The growth of plastics would be even faster if they weren't made out of such an expensive raw material, petroleum. But many of the materials they are replacing have important energy components in their creation as well. So the advances in plastics continue.

In this chapter, the big four thermoplastics are covered: polyethylene, polypropylene, polyvinyl chloride, and polystyrene. Like most other thermoplastics, they are long-chain polymers that become soft when heated and can be molded under pressure. They are linear- or branch-chained and, except for some exotic copolymers, have little or no cross-linking. Technological advances continue. Research in copolymerization, catalysts, processing, blending, and fabricating continues even as you read this.

Polyethylene

You have to shake your head in wonder when you think about how the largest selling plastic was developed—by accident. In 1933, the scientists at the ICI labs in England were attempting to make styrene by the high-pressure reaction of benzaldehyde with ethylene. Instead, they ended up with a reactor lined with a solid, white, wax-like material—polyethylene. (*See* Figure 23–1.) Six years later, a German scientist at IG Farben-Industrie, Max Fischer, was attempting to synthesize lube oils from ethylene. He tried a catalyst of aluminum powder and titanium at low pressures and ended up with a solid, white, wax-like material—polyethylene again.

$$\left[CH_2 - CH_2 \right]_n \qquad \text{Indicates repeating links}$$

Fig. 23–1 Polyethylene

The English experience eventually developed into the high-pressure polymerization route to Low Density Polyethylene (LDPE). The German experiment was the forerunner of the low-pressure route to High Density Polyethylene (HDPE).

The most recent arrival was Linear Low Density Polyethylene (LLDPE) in 1977. LLDPE combined some of the best features of both LPDE and HDPE by using a comonomer, butene-1, hexene-1, or octene-1.

While the ongoing story about new product development has unfolded, the parallel, more technical side of the saga has continued—catalysis, the science that made it all possible. In the early processes through the 1940s, the commercial polymerization processes used free radical, peroxide-base catalysts to kick off the polymerization reaction. The popular catalysts included t-butyl peroxypivalate, t-amyl peroxypivalate, t-butyl hydroperoxide, and t-butyl peroxybenzoate, chemicals you don't run across every day. These catalysts were typically used in high-pressure processes to produce branched LDPE.

The next wave came in the 1950s and 1960s when Karl Ziegler and Giulio Natta came up with their sensational metal-based catalyst systems. They derived their most widely used catalysts from titanium and organo-aluminum compounds such as titanium tetrachloride/trialkyl aluminum or other transition metals such as zirconium and vanadium in place of the titanium. Also during these decades, some commercial processes used catalysts based on chromium or molybdenum oxides supported on silica or silica alumina. The advantage of these catalysts came from their ability to catalyze ethylene into polymerizing at low temperatures and pressures, precipitating the commercialization of the solution, slurry, and gas phase processes to produce linear HPDE. More later.

The 1970s saw the introduction of higher activity catalysts based on magnesium chloride-supported titanium that improved the control of the physical properties of the polyethylene—molecular weights, stereospecificity, and the degree of copolymerization.

In the 1980s, researchers introduced bis-cyclopentadienyl zirconium dichloride, catalysts that maintained their high level of activity for long periods of time. These catalysts became the favorites for copolymerizing ethylene with alpha olefins to produce LLDPE.

The latest, but undoubtedly not the last, wave of innovative catalysis in the 1990s brought metallocenes to the scene. These catalysts are based on the usual zirconium, titanium, vanadium, or palladium metals but they have a mind-numbing atomic structures called coordination complexes which jump start the growth of the polymers. Bis-cyclopendienyl/nandium chloride is a typical metallocene. These catalysts are now used in all four processing routes. They have enabled new comonomer combinations, including polymerization with styrene, acrylates, carbon monoxide, vinyl chloride, and norbornene, a cyclic olefin that helps promote cross-linking.

Metallocenes give polyethylene producers a long list of opportunities to work on. They have already created polyethylene copolymers that compete well in applications that have been formerly the exclusive domain of the more costly, so-called high value plastics. Further, they are augmenting the chromium oxide and Ziegler-Natta catalysts systems that have been used for HDPE and LLDPE with metallocene catalysts. That creates even further control of properties at no additional capital and operating costs except for the catalysts.

Producers use four routes to make polyethylene, the bulk- or high-pressure process, the solution-phase process, the slurry-phase process, and the gas-phase process. Organizing your thinking around processes and products is not all that straightforward. Some of the processes can be used to produce all the polyethylene forms, some only a few or one. That calls for a few words first by product and then more by process.

The most important things about polymers is properties—how they look, how they react, and how they perform when you do things to them or with them. That's why different types of polyethylene were commercialized. What are the primary differences between LDPE and HDPE? LDPE is more flexible and has better clarity; HDPE has greater strength and less creep, and is less permeable to gases. That seems to go along with density differences, but also has to do with molecular weights, branching, and crystallinity, as you'll see below.

LLDPE has most of the good features of both LDPE and HDPE—strength, flexibility, clarity, good dielectrics, and high/low temperature stability (for wire and cable shielding).

Ironically, the high-pressure process produces a low-density product; low medium pressure produce a high-density material. You'd think it would be just the opposite. But it has to do with branching and crystallinity. The high-pressure leads to less crystalline molecules: the less crystalline, the less dense. (Recall the past example in Figure 22–6. Uncooked spaghetti is denser than uncooked macaroni, and the spaghetti-shaped polymers are completely crystalline.)

In any event, the difference in the densities of these polyolefins is small. For each of them, the density varies according to the degree of polymerization generated in the process. But in general, LDPE is about 0.920–0.935 grams per cubic centimeter; HDPE is about 9.955–0.970 g/cc; LLDPE varies between 0.920 and 0.950. That's a variation of less than 5%. So molecular weights and chemical structure also influence properties of the three different polyethylenes.

Bulk or high-pressure process

Based on the original commercial processes from the 1940s, this route runs at pressures as high as 50,000 psi and temperatures of 400–650°F. The process has its limitations since free radical peroxide-base catalysts cannot yield linear, highly crystalline polymers. They create branched chains. Today the process is used for LDPE.

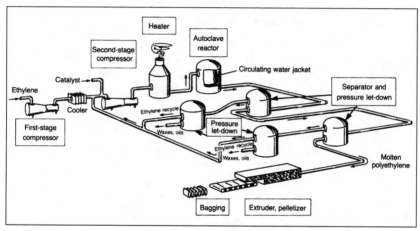

Fig. 23–2 High pressure process for LDPE

The process in Figure 23–2 shows the compression of the ethylene in two stages. (There are more.) Ethylene will start to polymerize on its own in an uncontrolled fashion at 212°F, so in between compressors, the gas needs to be cooled. (Compression always makes the gas temperature rise. That's why the bottom of your bicycle pump is hot after you've filled your tire.)

The compressed ethylene and a peroxide initiator (catalyst) enter the autoclave reactor. An autoclave is any vessel that can be closed up and maintain pressure at elevated temperatures. In chemical applications, the ongoing reaction inside the autoclave generates heat and/or pressure. (When a doctor or dentist sterilizes his instruments in his office, he uses an autoclave that generates superheated steam.)

The residence time of the ethylene and the initiator in the autoclave runs 30–120 seconds. Actually, the producer sets the reaction conditions, temperature, pressure, catalyst selection and concentration, and residence time to determine the physical properties of the polymer formed. The reaction is exothermic, and it is run under adiabatic conditions, meaning no heat has to be added. The temperature is maintained at a constant level in two ways to meet quality specifications. First, cooler ethylene is fed to the autoclave continuously and hotter polyethylene is drawn off. Second, the autoclave has a water jacket that operates as a big sponge, sopping up excess heat that variations in the reaction rate can cause. The water jacket also acts as insurance against a runaway exothermic reaction.

To separate any ethylene from the LDPE in the autoclave effluent, the pressure is let down in successive vessels and the ethylene flashes off (vaporizes). It is recycled to the compressors. The LDPE in a molten (hot liquid) state is cooled, extruded, pelletized, dried, and bagged.

Often mineral oil (a simple, medium-weight petroleum product) is used as a carrier for the catalyst. The amounts of both, very small in proportion to the LDPE, are generally left in the LDPE. The mineral oil ends up acting as a plasticizer.

Fig. 23-3 Celanese's HDPE plant

Solution phase process

Originally used to make HDPE (*see* the plant in Figure 23-3), this process has been adapted to copolymerization and the production of LLDPE. The system runs at much lower pressures, around 1200 psi, and temperatures of 400–600°F using the Ziegler-Natta catalysts. The metallocene catalysts improve the feed flexibility and range of properties. This process can also use ethylene with purity as low as 85%, and a wide range of comonomers, giving it an advantage over other routes.

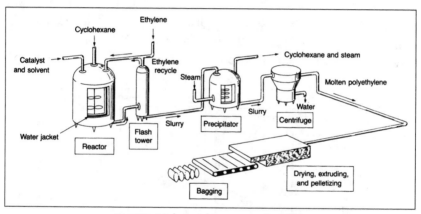

Fig. 23–4 Solution phase process for HDPE

In the process shown in Figure 23–4 the reactor is partially filled with cyclohexane, the medium in which the reaction takes place. The cyclohexane keeps the ethylene, the catalyst, and the polyethylene fluid and in contact with each other; it sponges up much of the heat from the exothermic reaction and helps control the rate of ethylene consumption.

The ethylene resides in the reactor about two minutes, and as it polymerizes, it remains dissolved in the cyclohexane. To keep the concentration of polyethylene in the cyclohexane at 35–40%, a solution of the feed, solvent, and product are continuously drawn off. Downstream, the ethylene flashes off to be recycled in a flash tower. A precipitator removes the polyethylene from the cyclohexane by centrifuge. The polyethylene is steam-stripped to remove any remaining cyclohexane, then dried, extruded, pelletized, and packaged.

The recycle streams, especially the cyclohexane that has come through the steam stripper must be thoroughly dried. It doesn't take much more than trace impurities to poison the fresh catalysts.

Slurry phase (or suspension) process. The unique-looking equipment in Figure 23–5 is a loop reactor. This process also takes place in a solvent (in this case, normal hexane, isobutane, or isopentane) so that the mixture can be pumped continuously in a loop while the polymerization is taking place. Feeds (the solvent, comonomer if any, ethylene and Ziegler-Natta catalyst) are pumped into the loop and circulated. Polymerization takes place continuously at temperatures below the melting point of the polyethylene allowing solid polymer particles to form enough to form slurry. The reaction takes place at 185–212°F and 75–150 psi. A slurry of HDPE in hexane is drawn off continuously or intermittently.

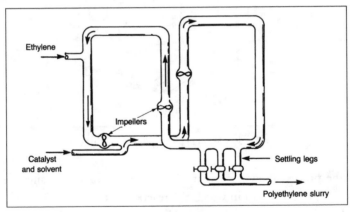

Fig. 23–5 Slurry process

The loops are pipes of 10- to 20-inch diameters, about 50 feet high, with a total length of 250–300 feet. They hold about 600 cubic feet of slurry and are water-jacketed to control the heat. The reaction temperature in the process is less than 212°F, with pressures of only a couple of hundred pounds, so the process is more economical (energy saving) than the others already discussed. After the slurry is withdrawn from the loop reactor, processing is the same as that downstream of the reactor in Figure 23–2. This process needs only small amounts of catalyst so catalyst separation from the reactor effluent is often not necessary.

Because of the longer residence times, this type reactor can make poly-ethylene molecules of higher molecular weights that also have high melt temperatures. Both HDPE and LLDPE are produced this way.

Gas phase process. This most widely used process yields both HDPE and LLDPE with a wide range of copolymers. Its simplicity begets its popularity. In addition, it accommodates a broad range of interesting property combi-nations used in both HDPE and LLDPE markets.

Popular comonomers include propylene, 4-ethyl-pentene (one of the forms of isohexene), normal hexene, and octane, all depending on the physical properties required of the polymer.

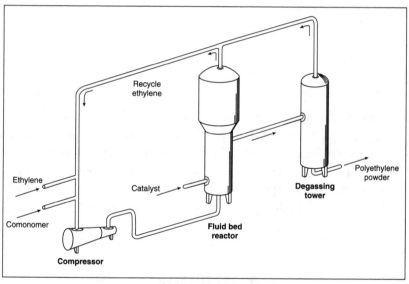

Fig. 23–6 Gas phase process for polyethylene

In Figure 23–6, polymer grade ethylene and any comonomers are blown into the base of a fluidized bed reactor. A very reactive catalyst (based on titanium and magnesium chlorides) is injected and admixes with the ethylene. Polymerization takes place at 150–212°F and 300 psi, and polymer particles stay in the fluidized state as the ethylene swirls through the reactor. Since the temperature is controlled at or below the melting point, the particles form a white powder.

At the top of the bottom section of the reactor, the polyethylene powder is separated from the unreacted ethylene and dispatched out the side of the reactor to a degassing tower. The ethylene continues to the top of the reactor, losing pressure and dropping out any remaining powder. In the degassing tower, any residual ethylene is recovered and recycled, together with the ethylene from the top of the reactor. The polyethylene powder is transported to the finishing area where it is extruded, pelletized, dried, and bagged.

Commercial aspects

Polyethylene in its various forms account for about 57% of the ethylene produced in the world, with the following distribution, more or less:

LDPE	–	32%
LLDPE	–	24%
HDPE	–	44%

The reason for this distribution is, of course, the market demand for physical properties, tempered by economics. While LLDPE has superior qualities in most aspects, HDPE and LDPE have old, mature, and in many cases, cheaper processing costs. Still, with LLDPE penetrating markets as traditional as trash bags, industrial liners, and injection-molded products, the displacement of LDPE and even some HDPE continues. Despite this, ever-larger plants of all four kinds are being built with capacities of 200,000–400,000 metric tons per year, scales intended to achieve even lower costs.

POLYPROPYLENE

When the polypropylene (PP) technology finally ripened in the late 1950s, the chemical industry was quick to harvest numerous applications. The primary attractions of this thermoplastic were the ease of molding or extruding it and its ability to hold color. Some of the familiar applications are automotive parts, luggage, pipe, bottles, fiber (particularly carpet face fiber and rope), housewares, and toys.

Some special problems arise in explaining PP, and they breed a set of new terms that are used throughout discussions on any polymer more complicated than polyethylene. The problem lies in the extra group that propylene carries along. Except for that methyl group, -CH$_3$, propylene (CH$_3$-CH=CH$_2$) would be ethylene.

The "allylic" hydrogens on the methyl group are reactive and capable of being displaced. (*Allylic*, from the Latin *allium*, meaning garlic.) The -C$_3$H$_5$ is the radical found in garlic and mustard. (*See* Figure 23–6.) That can lead to branching and sometimes cross-linking, which would affect the polymer's properties. As a matter of fact, the difficulty of controlling branching and cross-linking held up commercial development of PP until 1952. Then an Italian chemist, Giulio Natta, used Karl Ziegler's catalyst to produce a propylene polymer that finally had some useful properties.

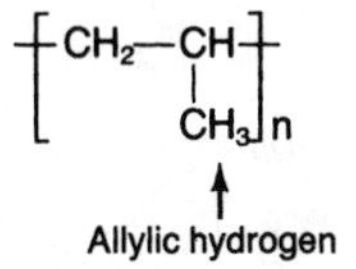

Fig. 23–7 Polypropylene

More importantly, understanding the chemistry of PP requires you to know the critical difference between PP and the polyethylenes—the asymmetry of the PP molecule's "backbone." In polyethylene, every carbon looks like every other carbon in the chain. In PP, the polymer linkage is between succeeding double-bonded carbons, like polyethylene. But, the methyl group survives as a branch on every second carbon in the PP "backbone" chain. (*See* Figure 23–7.) Furthermore, the orientation of that branch is crucial to the properties of the polymer. (*See* Figure 23–8.)

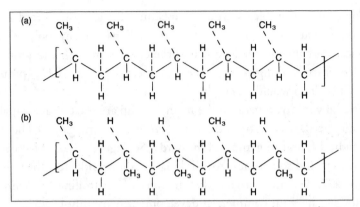

Fig. 23–8 (a) Isotactic polypropylene (methyl groups in the same plane);
(b) atactic polypropylene (methyl groups randomly in and out)

There's a whole area of chemistry dealing with the spatial configurations of organic molecules called stereochemistry. To get into this area, you have to have molecules that have an asymmetrical carbon atom. That's one that has four dissimilar atoms or groups attached to it. PP has that condition on a repeating basis—the methyl groups on every other "backbone" carbon. Such a polymer can be stereoregular or stereospecific.

In PP, stereoregularity of the methyl group is important. It really makes a difference whether every one sticks out in the same direction (more accurately, same plane). There are three possibilities that have been identified in PP molecules.

1. **Isotactic:** All the methyl groups are in the same plane
2. **Syndiotactic:** The methyl groups are alternately in the same plane
3. **Atactic:** The methyl groups are randomly in and out of the plane

Of these three PP isomers (called that because they all have the same formula, just different stereoconfigurations), isotactic makes the best plastic. Atactic polypropylene is soft, elastic, and rubbery but not as good as rubber, natural, or synthetic. It is usually separated from the isotactic propylene and discarded as waste, which adds considerable cost to the remaining isotactic. The isotactic form has a high degree of crystallinity with the chains packed closer together. (This makes sense since the molecules are more regularly oriented.) The greater crystallinity gives higher tensile strength, heat resistance, dimensional stability, hardness, and a higher melting point.

So you can see why branching in the polymerization process can be a problem—the symmetry is affected. And you can get a hint why PP was commercialized long after polyethylene. The chemistry and catalysis are a lot more demanding. That's why Giulio Natta won the Nobel Prize for his contribution to the field of stereo-catalysis, the discovery of the effects of titanium chloride and organo-aluminum compounds.

PP plants

The processes for making PP are very similar to those for polyethylene—bulk, slurry phase, and gas phase. In fact, the same plants shown in Figures 23–2, 23–4, and 24–5 can be used.

The slurry phase, the traditional route to PP, uses Ziegler-Natta type catalyst, a hydrocarbon solvent like hexane or heptane and polymer grade propylene (99.5%). Like the stringent requirements for polyethylene plant feeds, propylene must be high purity. Water, oxygen, carbon monoxide, or carbon dioxide will poison the catalyst. The reaction takes place in the liquid phase at 150–160°C and 100–400 psi. When the isotactic polymer particles form, they remain suspended in the diluent as slurry. The atactic polymers dissolve in the diluent.

After the polymerization step, the reaction mixture is fed to a heated separation tank where the unreacted propylene is flashed off and recycled. The polymer slurry is then washed with alcohol to deactivate and remove the catalyst and the atactic polymer (the bad stuff.) Centrifuging the slurry removes the diluent from the isotactic PP (the good stuff.) The product is washed with acetone, dried, and stabilized with suitable additives. It is sold as a powder or can be pelletized into granules.

In the bulk process, liquid propylene (polymer grade propylene here too) replaces the hydrocarbon diluent used in the slurry phase process. The PP is continuously withdrawn from the solution and any unreacted monomer is flashed off and recycled. The back end of the process, atactic PP removal and catalyst deactivation and removal, is the same as the slurry process.

Unlike the polyethylene counterparts, both these processes have to deal with the nearly 30% of the polymer coming out as atactic, limiting producers' enthusiasm for them. In the gas phase process, using the metallocene catalysts reduces the atactic propylene to less than 5%. In addition, feeding the lower costing chemical grade propylene is okay. As with the corresponding polyethylene process, the gas phase process for PP is carried out in a fluidized bed reactor using the titanium trichloride or magnesium chloride with a modified alkyl aluminum catalyst suspended in a solvent. The reaction takes place at 150–200°F and 350–500 psi.

Process improvements for all these processes are still underway, particularly with the use of metallocene catalysts. The primary objective is to get the percent of isotactic polypropylene to approach 100, minimizing the atactic form.

POLYVINYL CHLORIDE

Leaf and grass bags, upholstery, bottles, drainage pipe, roofing, flooring, coated fabrics, and siding are all made from polyvinyl chloride. PVC is *the* consumer product plastic.

Originally, vinyl chloride polymers were based on acetylene. The switch to ethylene chemistry came after the development of the oxychlorination process for vinyl chloride described in Chapter 9. Today very little acetylene-based vinyl chloride monomer (VCM) processing remains.

Vinyl chloride polymers and copolymers are often referred to as "vinyl resins." PVC is the most important member of the vinyl resin family, which includes polyvinyl acetate (PVAC), polyvinyl alcohol (PVA), polyvinylidene chloride (PVdC) and polyvinyl acetal. Almost always the term PVC includes polymers of VCM as well as copolymers that are mostly VCM.

Plasticizers

PVC (the homopolymer) is rarely used alone. Usually additives and plasticizers are added, more so than any of the other major thermoplastics. Plasticizers act like inter-molecule lubricants and can change pure PVC (and other polymers also) from a tough, horny, rigid material to a soft and rubber-like one.

Tricresyl phosphate (TCP, a gasoline additive in the 1960s alleged to be a tom cat exudate) used to be the popular plasticizer, but dioctyl phthalate has now replaced it. Dioctyl phthalate is the primary end-use for 2-ethylhexanol.

Plasticizer is usually added to the polymer during the compounding stage—that is, when it's being readied for molding, extruding, or rolling. The plasticizer is added in a hot mixer or roller operation. If PVC is expected to be plasticized, the polymerization steps can be controlled to produce a polymer particle that's very porous. Typically, for a flexible PVC, 25–30% of the finished PVC weight is plasticizer.

Adding plasticizer, like dioctyl phthalate, is generally accomplished by mechanical methods. Permanent or chemical plasticization can be done by copolymerization of VCM with monomers such as vinyl acetate, vinylidene chloride, methyl acrylate, or methyl methacrylate. Comonomer levels vary from 5–40%. The purpose of the co-polymers, of course, is to change the properties such as softening point, thermal stability, flexibility, tensile strength, and solubility.

Another way to vary PVC properties is to add in other polymers such as ABS, SAN, MMA, and nitrile rubber. These mixtures will improve the processibility and the impact resistance of the rigid PVC products.

Manufacturing PVC

Like polypropylene, PVC has the problem of stereospecificity. The carbon atom to which the chlorine atom is attached is asymmetrical. (*See* Figure 23–8.) As a result, PVC molecules can be isotactic, syndiotactic, and atactic. Commercial PVC is only 5–10% crystalline—low percent isotactic. It is more dense, 1.3 to 1.8 g/cc, than the polyolefins. (*See* Figure 23–9.)

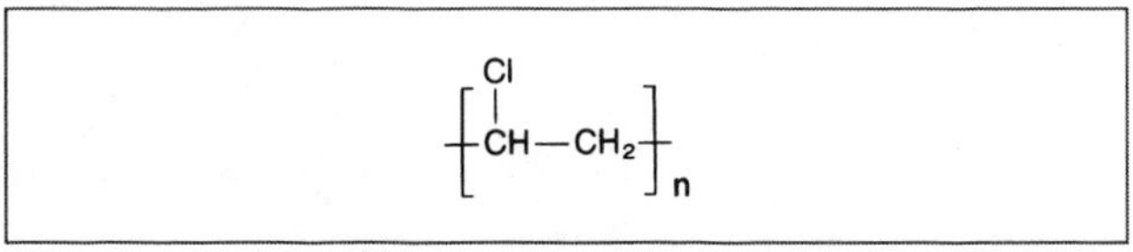

$$\left[\begin{array}{c} Cl \\ | \\ CH-CH_2 \end{array}\right]_n$$

Fig. 23–9 Polyvinyl chloride

VCM can be polymerized by all four processes: suspension, emulsion, bulk, and solution. Most PVC is made by the suspension method because the polymer is more suitable for molding, extruding, or calendering (that's calender, with an -er, which means rolling into thin sheets).

PVC in a latex form comes from the emulsion process, the second largest route. PVC latex can be used for coatings "as is" or can be made ready for molding. In that case, it is spray dried and the PVC particles are put into a liquid plasticizer (called plastisol) or into a mixture of plasticizer and organic solvent (organisol). The PVC particles do not dissolve but remain dispersed until the mixture is heated. Fusion then occurs, yielding the final plastic object. This is useful in forming special shapes by loading a mold with the plastisol or organisol and heating. (Vaporization of the organic solvent is often used to create foam. *See* Foams later in this chapter.)

The PVC plant

In the suspension polymerization process, the autoclave reactor is filled with water. PVA, polyvinyl alcohol is the dispersing agent that helps stabilize the suspension. Lauroyl peroxide is the free radical catalyst that starts it all off. The reaction temperature is around 130°F, and the process takes 10–12 hours per batch, with 95% conversion.

The reactors are typically 5000–6000 gallon, glass-lined, water-jacketed vessels. (*See* Figure 23–10.) After all the ingredients are loaded in, steam is run through the jacket to get the mixture up to 120–150°F. After the reaction begins, cooling water replaces the steam in the jacket to take away the heat generated in the exothermic process. Meanwhile, the vessel contents are mixed vigorously to keep the monomer suspended in the water. The PVA helps out here. The polymer molecules have to keep bumping into each other to keep growing.

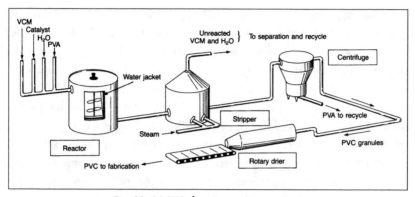

Fig. 23–10 PVC plant (suspension process)

To get a porous PVC bead that will accept high levels of plasticizer, a sudden pressure release is sometimes used. During the 10–12 hours of cooking, the reactor pressure will increase as the temperature goes up, then decline slowly as the polymerization approaches 100%. If the pressure is suddenly released during the process, some of the unreacted VCM will vaporize. The granules of PVC that have formed will begin to swell as it absorbs the VCM vapor. As the vessel is "buttoned back up" and repressured, the VCM will reliquefy, but the PVC already formed remains swollen and porous.

After the reaction is completed, the suspension is transferred to a degassing tank, where steam is used to strip out the unreacted VCM. The PVC, now in a slurry with PVA, is separated by centrifuging and drying. The PVC powder or granules are then ready for additives and plasticizers for fabrication into one of the three mediums in which it's used: calendered products, extrusion, and molded products.

POLYSTYRENE

When you hear polystyrene (PS) you probably think of products made of PS foam—disposable coffee cups, packing materials, buoys and boat bumpers, and cheap ice chests. As a matter of fact, PS foam products are so important that there's a special section at the end of this chapter dealing with foam.

Foam accounts for less than half of the polystyrene output. The remaining products have properties very different from foam. PS is an excellent plastic for molded automobile and refrigerator parts. It accepts color so well that it is widely used in molding applications to simulate wood. Probably the "wood" on your BMW dash is PS.

There's a lot of competition between PS and the other five big thermoplastics: LDPE, LLDPE, HDPE, PP, and PVC. Polystyrene continues to lose market share, but it seems to have a permanent place in some applications, particularly molded foams (for carryout food containers), some extrusions, and sheet and film applications. About half the polystyrene ends up in packaging, 17% in electrical/electronics applications, 13% in construction, building products and furniture, and 7% in medical applications.

Manufacturing PS

Like PP and PVC, each repeating monomer unit (*see* Figure 23–10) in PS has an asymmetric carbon atom. It's the phenyl group (benzene ring) attached to this carbon atom that makes the polymer asymmetrical. The polymer can be iso-, syndio-, or atactic. Commercially produced PS is usually an atactic amorphous polymer (low crystallinity, good optics). The isotactic form can be made using the Ziegler-type catalysts. However, there's no major, marketable improvement in its properties so most processes produce the cheaper atactic form. (*See* Figure 23–11.)

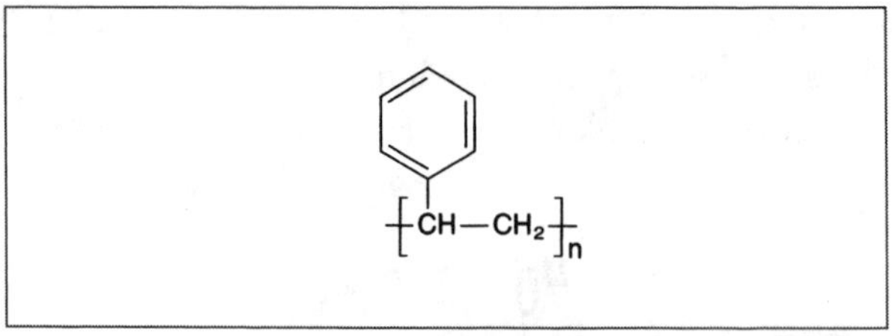

Fig. 23–11 Polystyrene

All four polymerization processes can be used to make PS. The reaction is an addition polymerization using a free radical initiator (benzoyl peroxide or di-tertiary butyl peroxide). Mostly, the suspension or bulk processes are used. The suspension process is identical to the PVC process shown in Figure 23–9. As just one more mind expander, the bulk process will be covered here in Figure 23–12.

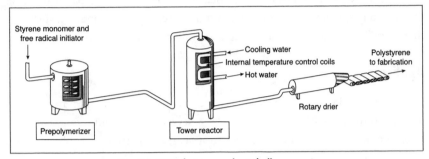

Fig. 23–12 Polystyrene plant (bulk process)

The polystyrene plant

The bulk polymerization process needs monomers that can dissolve their own polymers. (There's no solvent or water in the reactor to keep the polymer floating around.) Styrene and some of the more commonly used comonomers have this property, and so it's generally cheaper to use bulk polymerization.

The process begins in a prepolymerizer, which is a water-jacketed reactor with a mixer in it. (*See* Figure 23–12.) The styrene is partially polymerized by adding the peroxide initiator and heating to 240–250°F for about four hours. About 30% of the styrene polymerizes and the reactor contents become syrupy goo. That's about as far as the prepolymer step can go—30% conversion— because the mixing and heat transfer gets very inefficient as the goo gets thicker, and the polymerization becomes hard to control.

The goo is then pumped to the top of a vertical, jacketed tower with internal temperature-regulating coils. The vessel is kept full of the styrene/PS mixture. A temperature gradient (change) of 280°F at the top and 400°F at the bottom is maintained. The temperatures are controlled to prevent runaway, but to permit 95% conversion of styrene to PS. As the polystyrene molecules grow, they sink to the bottom of the vessel and can be drawn off. The residence time in this vessel is three to four hours. The molten PS is extruded to strands, chopped into pellets, and bagged.

The most critical factor in this process is temperature control in the second reactor. The viscosity of the mixture top to bottom changes with temperature, but also with PS concentration. If hot spots develop because of the exothermic reaction, a runaway can occur. In that event, the batch must be immediately quenched, ruining it. Several process improvements (not shown in Figure 23–12) include using agitators, solvents, and solvent removers.

Applications

About 25% of polymerized styrene is in the copolymer form. The largest volume copolymer is SBR (25% styrene, 75% butadiene-rubber), used for making tires, hoses, belts, footwear, foam rubber, rubber-coated fabrics, and adhesives.

ABS (30% acrylonitrile, 20% butadiene, and 50% styrene) is a tough plastic with outstanding mechanical properties. ABS is one of the few plastics that combines both toughness and hardness. So the applications include ballpoint pen shells, fishing boxes, extruded pipes, and space vehicle mechanical parts. There's more than 20 pounds of ABS molded parts in an automobile.

SAN (70% styrene, 30% acrylonitrile) has better heat and chemical resistance and is stiffer than PS. The optical clarity is not as good. SAN is used in a variety of houseware applications, particularly those things that will come in contact with food (chemical attack) and those that will end up in a dishwasher (heat attack). Coffee pots and throwaway tableware are good examples.

FOAMS

Foamed polymers are low-density, cellular materials that contain bubbles of gas and are made in a variety of ways out of thermoplastics and thermosets. Their properties vary from rigid to flexible. The rigid foams are best known for their insulation properties (like in ice chests). The flexible foams are used extensively in cushioning (seats, mattresses).

The difference between rigid and flexible foams, from a simplified view, is the nature of the cells that make up the foam. Rigid foams are made up of closed cells. The gas they contain is sealed in. Flexible foams have open cells. When you compress flexible foam, all the air can be squeezed out. When you compress rigid foam, the gas cannot escape, so nothing moves. Simple, isn't it?

The closed cells also give rigid foams their excellent insulating properties. Gas is a notoriously poor conductor of heat. That's why storm windows work. They have a dead air space in the middle. Rigid foams are just like storm windows. They trap a dead air space. Flexible foams wouldn't do quite as well because they let the air move around.

Foams are commercially produced several ways. Some polymerization processes produce their own foam. Polyurethanes, for example, are very exothermic. When they are formed, if a little water is present, CO_2 will be a by-product. As the polymer forms, the CO_2 will cause closed cell foam. As another example, a blowing agent can be injected into the molten polymer. The agent will later decompose, giving off a gas when the polymer is heated to melting. Epoxy resins are expanded into foams this way.

A popular, related technique is to inject some sort of a volatile material into the polymer while it's still molten, causing it to foam immediately. Fluorocarbons are used this way in making polyurethane foams suitable for insulation. The trapped fluorocarbon is even better than air as an insulator. Fluorocarbon or air is used to expand PS foams also. The gas is injected as the molten polymer is forced through a die. The foamed PS is then immediately injected into a mold to make items like egg cartons and trays for meats, produce, or fast-food.

Expandable PS beads are a material devised to accommodate the transportation drawbacks of foams. Foams take up a lot of room, but not much weight, so a truck or boxcar cannot be used very efficiently. Expandable PS beads can be readily turned into foam at their destination. The beads are impregnated with a volatile liquid like pentane as they are extruded, chopped, and cooled. Later, on site, the beads are heated in small batches with steam. The vaporization temperature of the pentane is just below the melting point of the PS beads. As the beads soften, the pentane flashes (volatilizes) and causes the PS to foam. The polymer is then ready for molding. Coffee cups, ice chests, life preservers, buoys, and floats are often fabricated this way.

Most thermoplastics and thermosets can be foamed, many of them into either flexible or rigid foams. The choice is controlled by the blowing agent, additives, surfactants, and mechanical handling. Some polymers can be expanded as much as 40 times their original density and still retain a substantial part of their strength. Most commercial foams are expanded to densities of two to five pounds per cubic foot. (Water is 62 pounds per cubic foot.)

PLASTIC PROPERTIES

That worn-out joke applies: "The three most important things about plastics are (1) properties, (2) properties, and (3) properties." It's impractical to cover all the dimensions, but here are some of the most important.

Polyethylene (PE) has excellent electrical properties, good clarity, good impact strength, and is translucent in thick sections. It also has good chemical resistance and excellent processibility.

PP is the lowest density plastic. It has fair-to-good impact strength and excellent colorability. It's translucent in thick sections, and it also has good chemical resistance. The properties can vary widely with different degrees of crystallinity. PP has good resistance to heat and low water absorption. That makes it a suitable material for many medical instruments that need sterilization by steaming.

PVC has good electrical properties and is flame-resistant with the proper plasticizers. It's even self-extinguishing. It has good impact strength and chemical resistance. Although rigid "as made," it is easy to make flexible by the addition of plasticizers. It does require heat and light stabilizer additives.

PS is easily processible and has excellent color, transparency, rigidity, dimensional stability, good tensile strength, and electrical properties. It is easily foamed.

Chapter 23 in a nutshell ...

The six most popular thermoplastics, low-density polyethylene, high-density polyethylene, linear low-density polyethylene, polypropylene, polystyrene, and polyvinyl chloride, are all linked in the same chemical configuration, via the ethylene group. The only difference is that three of them have substituted group for one of the hydrogens belonging to ethylene, $CH_2=CH_2$. Polypropylene has a methyl group hanging off, polystyrene has a benzene group, and polyvinyl chloride has chlorine.

The combination of these unique appendages—plus the different equipment, reaction pressures, and temperatures, use of comonomers and catalyst—result in quite different properties for each thermoplastic. But they all can be remolded, melted, or dissolved a number of times after initial formation.

EXERCISES

1. What does the term copolymerization mean, and give some examples?

2. What's the "bad" kind of polypropylene, and why is it "bad?"

3. What are all the building blocks and intermediate petrochemicals that lead up to polyvinyl chloride?

CHAPTER 24 ⬡

Resins and Fibers

"Tell her to make me a cambric shirt."

"Scarborough Fair" (1966)
Paul Simon and Art Garfunkel

After you've plowed through the thermoplastics, you only need to read about the resins and fibers to cover the rest of the applications for most petrochemicals. That's one reason for putting resins and fibers in one chapter. The other is that some polymers, like nylon, can be both a resin and a fiber. You just grow them a little differently.

The coverage in this chapter is compact—no detailed process descriptions or diagrams. Resins and fibers aren't really petrochemicals anyway. They're just a good climax to the petrochemical story.

RESINS

Chapter 22 didn't give you a very satisfying definition of resins. But it's useful here to talk about two classes of polymers called resins: thermosets and engineering thermoplastics.

Thermosets

You'll recall that thermosets are polymers that have lots of cross-linking. The molecules are three-dimensional, rather than two. More importantly, once the cross-linking bonds are in place, the polymer becomes rigid and hard. Put another way, once the thermoset occurs, it is irreversibly set. That's the difference between thermosets and thermoplastics. The latter can be remolded and reshaped; the former cannot. When you sweep up the scrap material around the molding/extruding machines that handle thermosets, you throw it away.

Phenolic resins. The oldest condensation reaction on record is between phenol and formaldehyde to produce phenolics. Professor Adolf von Baeyer first documented the reaction in 1872, for which the Nobel Committee awarded him their prize in 1905. Thirty years later, a technical application of this reaction was worked out by Dr. Leo Baekeland, when he showed that useful moldings can be made by carrying out the final stages of the reaction under pressure. As his reward, phenolic resins are still often called "Bakelite," a seemingly better deal than Baeyer's. At one time, phenolics were the work-horse of the plastics industry.

The chemistry of phenolic resins is complex. Even today it is not fully understood. The brief description that follows is definitely an oversimplification of the reactions involved.

To make phenolic resins, you have to first make a phenolic prepolymer, which may be in a liquid or solid form. This can be accomplished by using either a base or an acid catalyst. The prepolymer is a low molecular weight, linear polymer that—and this is the whole key to phenolics—can be further processed at a time of the processor's choosing to give the cross-linked phenolic resin. All you need is a little more heat and pressure. The reactions are shown in abbreviated form in Figure 24–1.

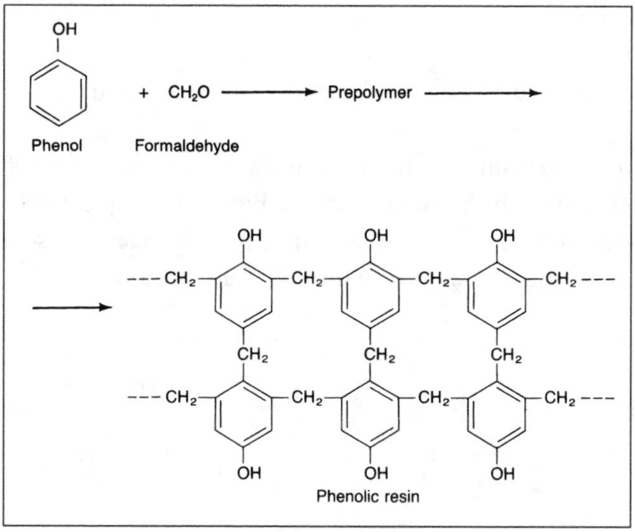

Fig. 24–1 Cross-linked phenolic resin

Phenolic resins are the cheapest of all molding materials, since they usually contain more than 50% filler—sawdust, glass fibers, oils, etc. Their main properties are heat resistance, excellent dielectrics, and ease of molding. However, they have poor impact resistance (they crack) and they don't hold most dyes very well, except black. Their use is thereby restricted—they're functional but not pretty. When the telephone companies started making phones in colors, they quit using phenolic resins and instead bought more expensive thermosets.

Other applications for phenolics are switchgears, handles, and appliance parts, such as washing machine agitators (that's why they're usually black). Phenolics are widely used to bond plywood, particularly exterior and marine grades. Although urea-formaldehyde resins are cheaper for this purpose, they were not nearly as water-resistant and have been limited to interior grades. Abrasive wheels and brake linings also are bonded with phenolic adhesives.

Epoxy resins. The reaction of Bisphenol A and epichlorohydrin gives a low molecular weight linear polymer. This polymer further reacts with an amine-curing agent, $R-NH_2$, to give a general-purpose thermoset. (*See* Figure 24–2.) Now that's not as complicated as it sounds. First of all, you'd think the name epoxy is used because the epoxy ring, $-C - C-$ first encountered in

$$O$$
$$/ \backslash$$
$$-C - C-$$
$$| \quad |$$

ethylene oxide, remains in the molecule. Well, it is and it isn't. It is in the prepolymer chain, which results from the Bisphenol A/epichlorohydrin condensation reaction. However, it is used up in the subsequent cross-linking step. This ring is the primary site where the polymer cross-links.

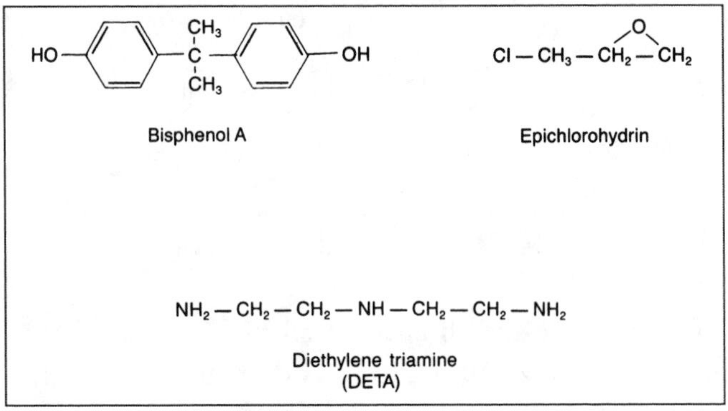

Fig. 24–2 Epoxy monomers

Epoxy resins are a post-World War II development. Initially, they were used as surface coatings. They are extremely resistant to heat and corrosion, and they have excellent adhesion to metals. Unfortunately, they get chalky when exposed to too much sunlight. Therefore, their use is limited to primers and places protected from sunlight, like coating pipeline interiors. Coatings are the largest volume use of epoxy resins, but the best known is the household adhesive that comes in two tubes. One contains the resin; the other has the "hardener," an amine catalyst plus filler. Common amine curing agents are diethylene triamine (DETA) and triethylene tetramine (TETA). The package is usually labeled "epoxy glue."

Other major uses are commercial adhesives, laminates, and potting for electrical components. Epoxy resins are particularly suitable for potting (imbedding electrical components in a nonconductive thermoset) because they provide dimensional stability (no shrinking) as the thermoset cures.

Polyurethanes. The word urethane has the same root word as urine, because both are related to urea, NH_2CONH_2. All three chemicals have the characteristic linkage, $-N-C-$.

$$\begin{matrix} | & || \\ H & O \end{matrix}$$

Urethanes result from reactions between alcohols and isocyanates. So what's an isocyanate? It's an organic compound having the $-N=C=O$ signature grouping. Isocyanates are the product of an amine and phosgene. An amine is a compound with the formula $R\text{-}NH_2$. Phosgene is $COCl_2$, a product resulting from the reaction of carbon monoxide and chlorine.

Simple polyurethanes are made from diisocyanates and diols. The general formula for diisocyanate is $O=C=N-R-N=C=O$. The popular one, toluene diisocyanate (TDI), is shown in Figure 24–3. Diols are molecules that have two hydroxyl groups (alcohol signatures) attached, such as propylene glycol. Polymerizing TDI and propylene glycol gives a linear polymer because both molecules are bifunctional. That is, they each have only two sites where they can react, so they string out end-to-end.

If cross-linking is desired, as in polyurethane foam, TDI can be reacted with a polyol or pololether. These chemicals have multiple -OH groups that make them polyfunctional. Cross-linking can take place at one or more of the hydroxyl sites. Compounds other than TDI can be used to make polyurethanes. However, TDI accounts for about 60% of this production.

Polyurethanes are primarily found in flexible foams, but they also make rigid foams and laminates. Two methods are used to develop these foams. In one procedure, all of the ingredients are mixed, including a catalyst and a blowing agent additive, and put into a mold. The reaction kicks off immediately. Foaming is completed in a few minutes because the gases are liberated from the blowing agent. However, complete curing (cross-linking) takes several more hours.

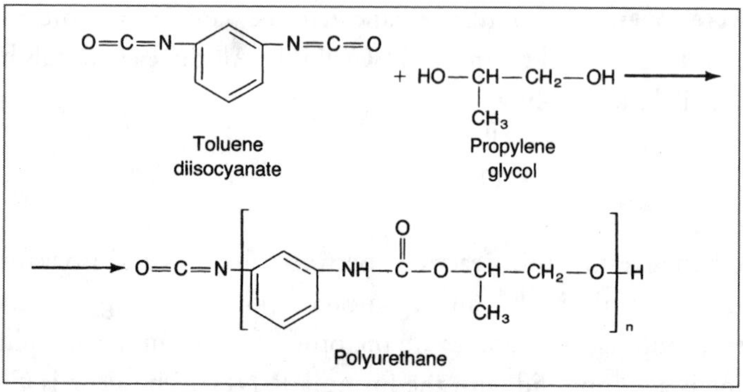

Fig. 24–3 Simple polyurethane

The other foaming process is referred to as the prepolymer method. The monomers are reacted to form a low molecular-weight prepolymer. Later the prepolymer is mixed with small amounts of water and is heated. The water reacts with the free isocyanate groups to liberate carbon dioxide, which foams the polyurethane as cross-linking starts.

Polyurethane foams are lighter than foam rubber and have displaced many of its applications, such as in bedding, cushions, car seats, armrests, and crash pads. Laminates are used widely in clothing as padding.

Rigid foams are excellent insulators, even better than polystyrene, and are used in refrigerators and refrigerated trucks and box cars. Polyurethane coating materials are popular additives to marine finishes and varnishes, particularly for gymnasium floors, bar tops, and other surfaces that take an abusive, abrasive beating.

Amino resins. Urea, the first recorded, synthetically produced, organic compound, can be reacted with formaldehyde to form polymers called urea-formaldehyde resins or amino resins. Its chemistry is similar to the phenolic resins.

Articles made from amino resins are water clear, hard, and strong, but they can crack. They have good electrical properties, and they have better colorability than phenolic resins. Amino resins are used as adhesives for plywood and particleboard but only in interior grades. They have low weather resistance and deteriorate when exposed to sun, heat, cold, and moisture.

Extensive use of these resins is found in textile and paper-treating and surface coatings. Many types of clothing also can be given a permanent press by an amino resin treatment. Amino resins can be molded and are used for radio cabinets, buttons, switch plate covers, dishware, and Formica.

Engineering resins

Engineers like to design machines that take plastics and inject them into molds, extrude them into filament, sheets, rods, and film, or blow-mold them into shapes. They want these plastics to have a terrific balance of mechanical, electrical, and chemical properties. That's why they have attached their name to engineering resins, a set of plastics unlike thermosets that they can remelt and further process mechanically. Nylon, polycarbonates, and polyesters are the three most popular engineering resins.

Nylon. The name *nylon* covers a number of polymer compounds, all of which are based on the amide linkage, $-\underset{\underset{O}{\|}}{C}-\underset{\underset{H}{|}}{N}-$.

They are shown in Figure 24–4. (Fibers made from nylon actually account for much more volume than resins made from nylon.) The two most popular nylons, both in resins and fibers, are Nylon 6 and 66. These two account for about 80% of the nylon production.

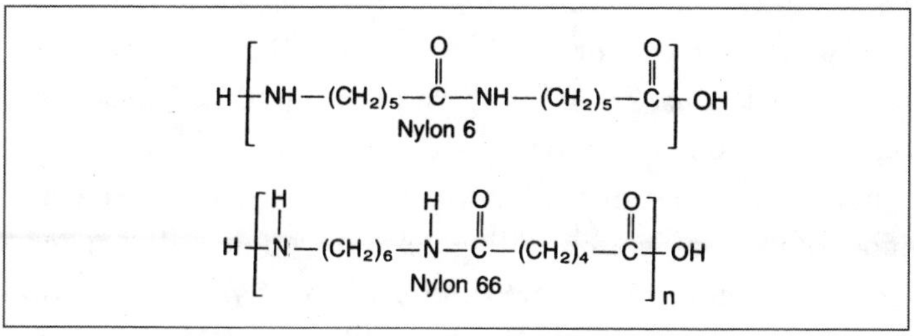

$$H-\left[NH-(CH_2)_5-\underset{\underset{O}{\|}}{C}-NH-(CH_2)_5-\underset{\underset{O}{\|}}{C}\right]-OH$$
Nylon 6

$$H-\left[\underset{\underset{H}{|}}{N}-(CH_2)_6-\underset{\underset{H}{|}}{N}-\underset{\underset{O}{\|}}{C}-(CH_2)_4-\underset{\underset{O}{\|}}{C}\right]_n-OH$$
Nylon 66

Fig. 24–4 Nylon

Nylon 6 is made by the addition polymerization of caprolactam. Caprolactam is a seven-sided heterocyclic that should be drawn as a septagon, just like cyclohexane and benzene are hexagons. However, hardly anybody can draw septagons well, so most people sketch caprolactam as a rectangle, like the one shown in Figure 24–5. The primary route to caprolactam is a complex, but 50-year-old process starting with cyclohexane. Three reactions later, out comes caprolactam. More recent process developments start with butadiene, a more efficient route that will probably replace the cyclohexane starting point.

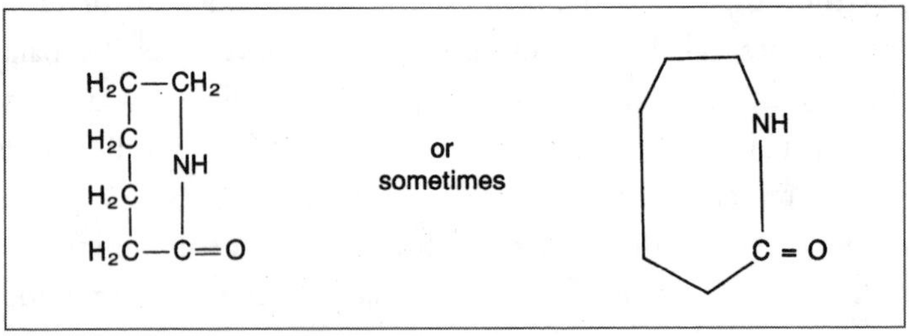

Fig. 24–5 Caprolactam

Nylon 66 is made by the condensation polymerization of adipic acid and hexamethylenediamine (HMD). Adipic acid was covered in Chapter 16 (cyclohexane is also the building block for adipic), and HMD shouldn't be such a threatening word to you by now. The hex is six; the methylene is $-CH_2-$. The di is two, and amine is the signature group $-NH_2$. Put them all together and they don't spell mother—they spell $H_2N(CH_2)_6NH_2$, which is HMD. The routes to HMD and adipic acid are shown in Figure 24–6.

The 6 and 66 are part of an awkward numbering system used to indicate what each nylon was made from. A number up to 12 indicates the nylon was made from a single monomer with that number of carbons in it. A number more than 12 signifies that two different monomers were used, with the number of carbons in each adding up to the number designated. Check Figure 24–4 to see how it works. Examples of other nylons are Nylon 11, 12, and 610. Originally, all double numbered nylons had a comma in the name—Nylon 6, 6. Through constant misuse—you know how careless those engineers are—the comma has mostly been dropped.

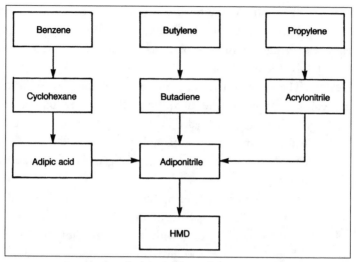

Fig. 24–6 Methods to produce adipic acid and HMD

Polycarbonates. The polycarbonates surfaced in the 1950s, so they are middle-aged polymers. They are made in a condensation polymerization process. The reactants are either Bisphenol A and phosgene or Bisphenol A, phosgene, and phenol. Since Bisphenol A is a derivative of phenol, the building block is the same in either case—phenol. The polycarbonate based on Bisphenol A has the best balance of properties. If you look hard, you can see the monomers in Figure 24–7.

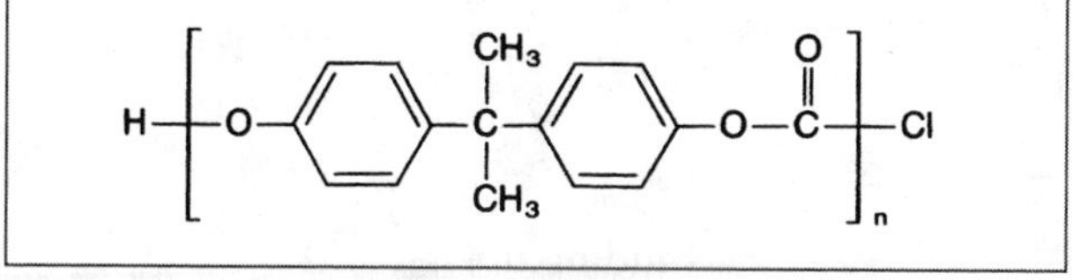

Fig. 24–7 Polycarbonate

Polycarbonates differ mechanically from the epoxy resins they resemble chemically because polycarbonates are plastics. They can be molded or extruded from chips or crumbs. The products have high impact strength and can be used to make sturdy films and transparent forms. That makes polycarbonates excellent candidates for electrical, electronic, and automotive applications and glass substitutes. Products include photographic film, goggles, helmets, streetlight fixtures, tool, machine, and appliance casings, and automobile taillight lenses.

Polyesters. The healthiest trend coming out of the plastics industry— personal health, that is—is the ubiquitous PET water bottle. Medical doctors all agree that the increased daily consumption of water enhances well-being. The improvement in producing PET, polyethylene terephthalate, in the 1970s, led to this pervasive replacement of glass bottles because of the cost and portability.

Thermoplastic polyester engineering resins come from an esterification polymerization process—terephthalic acid and ethylene glycol to give PET and dimethyl terephthalate reacting with 1,4–butanediol to give the less well-known PBT, polybutylene terephthalate. A simplified version of the PET reaction is shown in Figure 24–8. The process takes place in two steps, first making the relatively low molecular weight polyester prepolymer. The second step, done when the resin or fiber is about to be fabricated, is to lengthen the prepolymer by heating it to 525°F under a vacuum, producing the final polyester. Like Nylon, PET can be extruded or cooled and cut into chips for later melting.

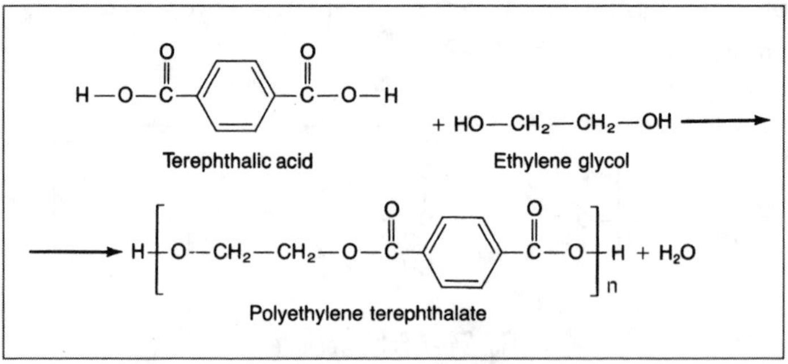

Fig. 24–8 Route to polyester

In 1999 Shell introduced their Corterra® polymers, polyesters made from 1,3-propanediol and terephthalic acid. This polymer, PTT (polytrimethylene terephthalate), was first synthesized in 1941, saw a long gestation period before someone found an economical production route to the propanediol glycol. PTT monomer will compete with Nylon in the engineering resins and fiber markets, especially carpeting.

Most of the polyester produced in the United States ends up in fiber applications—see the Fibers section below—but the growth of bottling applications, especially for beer, could change the percentages.

FIBERS

Natural fibers go back to prehistoric days. Probably one of the early applications was the conversion of a fiber (possibly wool or cellulose) into thread or rope strong enough to be used in a snare, net, or cage. Literature as far back as the 17th century notes that people attempted to make fibers out of something other than cotton, wool, or flax. The first man-made fiber, known as artificial silk, was made in the 19th century, when wood pulp was treated with nitric acid. The result was known chemically as cellulose nitrate and (eventually) commercially as Rayon. The commercial name referred to the sheen that "has the brilliance of the sun."

However, Rayon isn't a petrochemical derivative. Those products did not emerge until the technological breakthrough of W. H. Carothers of DuPont in the 1930s. His classical research showed that chemicals of low molecular weight could be reacted to form polymers of high molecular weight. Some people consider Carothers the father of polymer chemistry. His original offspring was Nylon 66, the first commercial fiber to be made entirely from a synthetic polymer. (Some reports have PVC fiber beating Nylon 66 by two years, but PVC never really caught on as a fiber.)

The first Nylon 66 fiber production facility came on stream in 1939. The first nylon stockings were marketed in May 1940—just in time for G.I.'s to take them to Europe in World War II.

DuPont continued their leadership role in synthetic fibers by commercializing acrylic fibers (Orlon) in 1950. They did a repeat performance in 1953 with a polyester called Dacron. The big four fibers—Nylon 6, Nylon 66, acrylics, and polyester—now account for most of the synthetic production and about half of the fiber production of all kinds, including cotton, silk, and wool.

The mechanics of fibers

The chemistry of fibers is the same as that for resins. The important difference is the mechanics. For polymers to be suitable for fibers, you must be able to draw them into a fibrous form, normally by extrusion. Second, the size and shape of the molecules that make up the fiber must be correct. To have acceptable fiber properties, the molecules must be long, so they can be oriented to lie parallel to the axis of the fiber. Normally, that's done (or enhanced) by drawing or stretching the fiber to several times its original length. The essential differences then between resins and fibers are the shape and the orientation of the molecules.

Spinning is the classic process of twisting a bundle of parallel short pieces of natural fiber into thread. (You can't get more classic than a spinning wheel.) It also is the modern process for extruding long lengths of man-made fibers. In the classic process, the short lengths are known as staple fibers, and the resulting thread is a spun yarn. The long, extruded lengths are called continuous filament or just filament yarn.

The denier of a fiber or a yarn defines its linear density. This term describes the weight in grams of a 9000-meter length of fiber at 70°F and 65% relative humidity. Denier generally denotes the size or diameter of the filament or yarn. Fibers usually range from 1–15 denier and yarns 15–1650 denier. Breaking tenacity is a measure of strength of a fiber or yarn and is measured in grams per denier. The stronger the fiber is, the higher the tenacity.

Nylon (polyamide fibers). The chemical structure of the nylon fiber looks just like the nylon resin. The polymerization processes are the same; the numbering systems are the same; and the two most important nylon fibers are the same nylon, 6 and 66. The difference is the length of the molecule in comparison to the cross-section. That's regulated by the polymerization process conditions.

After the polymer is formed, the filament is produced by extrusion. The nylon can be taken directly from the polymerization process while it is still molten. Alternatively, previously dried nylon chips can be heated and melted. The molten nylon is pressed through a spinneret, which is like a showerhead with 10–100 holes. (A spinneret with a single hole produces a monofilament. Nylon monofilaments are used for fishing line and sheer hosiery.) The fine strands that form are immediately air-cooled, stretched, and wound on bobbins.

Then the fibers undergo cold drawing. Nylon 66 can be stretched 400–600% of its original length with very little effort. A very important process goes on during the stretch. The polymer molecules orient themselves to a controlled amount of crystallization, depending on the stretch conditions. This crystallization gives the fibers their special properties.

Finally, the fibers are textured for specific applications. They can be twisted, coiled, or even randomly kinked if they are to be used for carpet piling. More than half of the total carpet-fiber market is based on nylon staple and filaments. When nylon is to be blended with other fibers, the filament is cut into staple fiber—short pieces 3–15 centimeters in length—for subsequent spinning.

One of the larger uses of nylon fibers is tire cord. In apparel applications, which are another major area, permanent press can be achieved by heat treatment. This crease resistance lasts until abrasion, heat, or pressure wears down the molecule orientation. Since it is strong and lightweight, nylon also is used for rope, parachutes, and some undergarments.

Polyester fibers. The process for producing polyesters was covered lightly in the sections on engineering resins. Read it carefully, and it will make sense, despite the formidable nomenclature and diagrams. It's really not that hard.

Polyester is the Madison Avenue name for PET, polyethylene terephthalate. Be careful, though, because PET is not a polyethylene-type chemical—it's a xylene derivative. If you remember that esters usually end in -*ate* and are based on the acid from which they are derived, then you can see that the "ester" in polyester is ethylene terephthalate.

Like nylon, when it is made PET can be extruded, cooled, cut into chips, and stored for later melt spinning. It also can go directly to the spinneret. The downstream operations of the spinneret are much the same as with nylon.

Unlike nylon, which is highly crystalline, PET fibers are amorphous after spinning. They are like the molecules shown at the top of Figure 22–6 in Chapter 22. In order to make a usable textile yarn or staple fiber out of PET, it must be drawn under conditions that result in orientation and crystallinity. This is accomplished by drawing at temperatures of about 175°F with stretch 300–400%. As with nylon, the conditions of draw (especially the amount) determine the tensile strength and shrinkage properties. Industrial PET fibers, such as those suitable for tires, are more highly drawn.

Other important properties of PET fibers are heat setting (permanent press) and the ability to blend with cotton and wool. By appropriate texturing, various finishes are possible: fur-like deep pile for coats, jackets, bath mats, soft toys, and strong, coarse fiber for tire cords, V-belts, fire hoses, and carpeting. Polyesters are very resistant to degradation by sunlight. Dacron is still the most popular sailcloth because of its weight, its quick drying characteristic, its resistance to stretch and to mold, and its color fastness.

Acrylic fibers. Acrylic fibers are polymers of acrylonitrile and another chemical. When acrylonitrile is 85% or more of the polymer, the fiber is called acrylic. If there's more copolymer so the percentage of acrylonitrile decreases to 35–85%, the fiber is called modacrylic. Some of the popular monomers used as copolymers are methyl acrylate and methacrylate, acrylamide, vinyl acetate, vinylidene chloride, and vinyl chloride. Dynel is 40% acrylo and 60% vinyl chloride.

These polymers are generally made by the solution polymerization method. The cooking time in the reactor takes 30–60 minutes before the chains are long enough. The basic polyacrylonitrile chain is like the polypropylene chain—it grows along the ethylene backbone. The cyanide grouping is hung over the side.

Melt spinning is not used for polyacrylics because they are sensitive to high temperatures. They actually begin to decompose before they reach melting temperature. Solution spinning is used instead. The dried polymer is dissolved in a polar solvent like acetone or dimethyl formamide (DMF). The spinning mechanics are otherwise the same, except the solvent is recovered as it vaporizes, immediately after the extrusion through the spinneret. Most acrylics are sold and used in the form of staple fiber.

Acrylic fibers generally have good "hand," as it's called in the business (they're soft). They resist creasing, and they are quick to dry. Acrylics have replaced wool in many applications, such as blankets and sweaters. Because of their unique bulking characteristics, they take on the appearance of wool yarn.

Polypropylene fibers. A small part of the total fibers market (and therefore at the tail end of this section on fibers) is fiber grade polypropylene. The chemistry for polypropylene fibers is the same as for thermoplastics. The spinning mechanics are the same as that for nylon. Polypropylene fibers are particularly resistant to abrasion and chemicals, and they are lightweight. However, they don't take colors very well, and the materials have low softening points and low resilience (they wrinkle). The major applications for polypropylene fibers are carpet-face fiber and backing (because it's tough) and rope (because it is strong and floats in water).

Chapter 24 in a nutshell ...

Thermosets are strings of several monomers, polymerized into high molecular weight compounds and cross-linked across the strings by chemical bonds. The cross-linking is what prevents the remolding, melting, or dissolving. Phenolics, epoxies, and polyurethanes are examples.

Phenolic resins are based on phenol and some other comonomer like formaldehyde. Phenolics are widely used in areas of tough duty, and where they don't have to look pretty, such as switchgear, handles, and plywood glue.

Epoxies are made from Bisphenol A and epichlorohydrin. They are tough, resistant to heat, and have good adhesive properties. They can be mixed on-site by adding a catalyst and filler to the already linked Bisphenol A/epichlorohydrin that causes the cross-linking and the "set" in thermoset.

Polyurethanes can be made from toluene diisocyanate and propylene glycol, or more complicated forms of these compounds. Polyurethanes are easy to foam either in flexible or rigid form.

Engineering resins are polymers designed to excel in certain physical characteristics. For the most part they are thermoplastics, but they can be thermosets as well. Examples are Nylon 6, Nylon 66, polycarbonates, and polyesters. Phenolics can be considered engineering resins.

Nylon 6 is made by polymerizing caprolactam, a derivative of cyclohexane. Nylon 66 is made by the polymerization of adipic acid and hexamethylenediamine. Polycarbonates are made from Bisphenol A and phosgene. Polyesters result from the reaction of terephthalic acid or its methyl ester with a glycol monomer.

Fibers, like Nylon 6, Nylon 66, acrylics, and polyesters, have to be made from thermoplastics because making the filaments requires that the molecules be oriented (literally all be lined up in the same direction). The orientation happens when the thermoplastic is reheated or dissolved in a solvent and squeezed out like spaghetti.

EXERCISES

1. What are the ingredients of polyurethane? Is polyurethane a thermo-plastic or a thermoset?

2. What's the difference between rigid foams and flexible foams?

3. Which of these are thermosets?

    nylon
    polycarbonates
    bakelite
    epoxides
    polyesters

COMMENTARY 4 ⬡

"Now you can't break
the ties that bind."

"The Ties That Bind"
Bruce Springsteen, 1949–

REVIEW

How do you summarize a summary of a whole body of chemistry like the polymers? You only pick out some tidbits, and you can summarize "what comes from what" with the following table.

The differences in the properties of polymers go back to the chemical configurations. In simple terms, thermoplastics can be molded because they are long-chain molecules that slip if pushed or pulled, especially at higher temperatures. Thermosets are cross-linked, so the long chains stay put under stress, strain, or heat. They don't melt, and they can't be molded once they set. The fibers get their flexibility and strength when the polymer molecules align during filament formation.

If you've gotten this far (even if you've skipped a couple of pages or chapters), you need to be reminded of the good news. There's a prodigious glossary and a wondrously complete index coming up right after the Appendix. You can use them to refresh your memory on those one or two points that you might forget sometime in the future.

PLASTICS AND RESINS COMPOSITIONS

Thermoplastics

Abbreviation/Name	Monomer
LDPE - low density polyethylene	ethylene
HDPE - high density polyethylene	ethylene
LLDPE - linear low density polyethylene	ethylene and butene-1 or other low molecular weight alpha olefins
PP - polypropylene	propylene
PVC - polyvinyl chloride	vinyl chloride
PS - polystyrene	styrene

Thermosets

Phenolics	phenol and formaldehyde
Polyurethane	diisocynate and propylene glycol or a polyglycol
Epoxy resin	Bisphenol A and epichlorohydrin
Urea-formaldehyde	Urea and formaldehyde
Polycarbonate	Bisphenol A and Phosgene

Fibers

Nylon 6	Caprolactam (from cyclohexane)
Nylon 66	Adipic acid and hexamethylenediamine
Polyacrylics	Acrilonitrile
PET - polyester	Terephthalic acid and ethylene glycol

APPENDIX 1 ⬡━━━━━

Conversion and Yield

"Work for the work's sake,
then, and it may be
That these things shall be
Added unto thee."

Kenyon Cox, 1856–1919

Conversion and yield are often used to describe the efficiency of a plant or a process. Sometimes the semantics get mixed up, and one is substituted (erroneously) for the other.

Conversion is defined as the percent of the feed that disappears in a chemical reaction. Take the pounds of benzene, for example, coming out of the ethylbenzene reactor, divide by the pounds of benzene going into the reactor, subtract that from 1.0 and multiply by 100, and you get percent benzene conversion. Conversion is usually measured around a reactor, not around the whole plant. In the case of the ethylbenzene plant, since there is a benzene recycle, there is virtually no benzene leaving the EB plant. So plant conversion is often not a very helpful concept, but the conversion across the reactor is.

Yield is a more difficult concept, first because it's used in several different ways, and second because one of the uses requires some basic chemistry. In the simplest case, yield refers to the pounds of product leaving a reactor divided by the pounds of feed. In an olefins plant, for example, the ethylene yield is equal to the pounds of ethylene divided by the pounds of, say, gas oil feed. (This definition of yield is more commonly used in refining than in petrochemicals, but then, ethylene plants are usually in refineries).

The petrochemical and more technical definition of yield refers to the amount of converted feed that actually ends up as product. This is not an easy subject for "nontechnical" discussion, but here goes.

DEFINITIONS

Atomic weight—the relative weights of atoms; hydrogen is the lightest and is defined to be one; carbon is 12 and oxygen is 16, meaning they are 12 and 16 times as heavy as hydrogen.

Molecular weight—the sum of the (relative) weights of atoms that make up a molecule; the molecular weight of water, H_2O is $2 + 16 = 18$; of methane, CH_4, is $12 + 4 = 16$.

Mole—The molecular weight of a compound expressed in grams is the gram molecular weight, usually contracted to the term mole. If the weight is expressed in kilograms or pounds, then it's kilogram moles or pound moles. A mole of water would be 18 grams of water; a pound mole of water would be 18 pounds of water.

The theoretical yield of a reaction can be calculated based on molecular weights. For example, the reaction of hydrogen with carbon dioxide to give methanol:

$$3H_2 \quad + \quad CO_2 \quad \rightarrow \quad CH_3OH \quad + \quad H_2O$$
$$3(1+1) \quad (12+16+16) \quad (12+1+1+1+16+1) \quad (1+1+16)$$
$$\text{or} \quad 6 \quad + \quad 44 \quad \quad 32 \quad \quad 18$$

Theoretically, three moles of hydrogen (6 grams) will react with one mole of carbon dioxide (44 grams) to give one mole of methanol (32 grams) and one mole of water (18 grams). In actual practice, a by-product, dimethyl ether, also gets formed:

$$2CH_3OH \rightarrow CH_3OCH_3 + H_2O$$

Suppose that in an experiment, 22 grams of CO_2 reacted with excess H_2O to give 12 grams of methanol and some dimethyl ether. To determine the *actual yield* of methanol based on the CO_2 reacted:

Step 1—Determine the moles of CO_2 reacted and the moles of CH_3OH produced:
22 grams of CO_2 is half of a mole of CO_2 (22/44) and 12 grams of CH_3OH is 0.375 moles (12/32)

Step 2—Theoretically, one mole of CO_2 will give one mole of CH_3OH.

Step 3—Divide the actual moles of product by the theoretical moles of product produced.

$$0.375 / 0.500 = 0.75$$
The actual yield is 75%.

Note that the yield is not the weight of the product (12 grams) divided by the weight of the feed (22 grams). Using molecular weights as an adjustment helps track what happens chemically to the molecules so that the chemist and the chemical engineer can determine whether to work together on improving the process.

APPENDIX 2 ⬡

Abbreviations of the Chemicals Industry

"This is the short and long of it."

**The Merry Wives of Windsor
William Shakespeare, 1564–1616**

AAAS	American Association for the Advancement of Science
AATCC	American Association of Textile Chemists and Colorists
ACS	American Chemical Society
ACS	American Chemist Society
AIC	American Institute of Chemists
AIChE	American Institute of Chemical Engineers
ANSI	American National Standards Institute
AOAC	Association of Official Analytical Chemists
AOCS	American Oil Chemists Society
API	American Petroleum Institute
ASM	American Society for Metals
ASTM	American Society for Testing Materials
C&EN	Chemical & Engineering News
CIIT	Chemical Industry Institute of Technology
CMA	Chemical Manufacturers Association

CMRA	Chemical Marketing Research Association
DOE	Department of Energy
DOT	Department of Transportation
EPA	Environmental Protection Agency
FDA	Food and Drug Administration
FTC	Federal Trade Commission
ICC	Interstate Commerce Commission
IUPAC	International Union of Pure and Applied Chemistry
NACA	National Agricultural Chemicals Association
NCI	National Cancer Institute
NF	National Formulary
NFPA	National Fire Protection Association
NIH	National Institutes of Health
NIOSH	National Institute for Occupational Safety and Health
NRC	National Regulatory Commission
NRDC	National Resources Defense Council
OSHA	Occupational Safety and Health Administration
SDA	Soap and Detergent Association
SOCMA	Synthetic Organic Chemical Manufacturers Association
SPE	Society of the Plastics Industry
TAPPI	Technical Association of the Pulp and Paper Industry
TSCA	Toxic Substances Control Act
USDA	U. S. Department of Agriculture
USP	U. S. Pharmacopoeia

GLOSSARY ⬦

"To find its meaning is my meat and drink."

Fra Lippo Lippi
Robert Browning, 1812–1889

A

Abherents. Compounds capable of promoting separation of two substances such as plastic or rubber products from the metal molds in which they are formed. Common release agents are waxy or fatty-like materials such as wax, tallow, or vegetable oil.

Absolute. Pure and free from admixture with other substances. (Absolute alcohol is ethyl alcohol dehydrated to 99% pure.)

Absolute zero. Minus 273°C or minus 460°F or 0°K or Kelvin, the scale used in theoretical physics and chemistry. Absolute zero is the theoretical temperature at which all molecular activity ceases. In practical terms, the lowest reachable temperature is about 1°K.

Absorbent. A solvent used to remove or extract one of the components of a mixed stream. The component is subsequently separated from the absorbent, usually by distillation. *See* absorption (with a *p*, not a *b*.).

Absorption. In theory, the penetration of one substance into another. In refining and petrochemicals processing, separation of gases by a scrubbing or washing operation with a liquid. Preferential solubility is shown for one or more of the components in the gas mixture allowing a separation process to occur. The liquid stream and dissolved gas(es) then undergo a desorption process that is basically the reverse of absorption—dissolved gases are freed. The liquid is then recycled back to the absorption tower.

Accelerator. A compound, usually organic, that greatly reduces the time for a reaction (usually a polymerization or vulcanization) to take place.

Acetylation. The reaction involving the introduction of the acetyl group (CH_3CO-). Common acetylating agents are acetyl chloride and acetic anhydride.

Acid. One of a large class of organic and inorganic compounds, all of which contain hydrogen. In water solutions, acid molecules give up one or more hydrogen ions, making them useful for a wide range of reactions. Inorganic acid examples are sulfuric (H_2SO_4) and Hydrochloric (HCl) acids; Organic acids include acetic (CH_3COOH) and adipic ($(CH_2)_4(COOH)_2$) acid.

Acid number. A measure of the amount of free acid in a substance equal to the amount of potassium hydroxide to neutralize it, in milligrams of KOH to neutralize 1 gram of acid.

Acidity. How far down the pH scale a substance is. *See* pH.

Activated alumina. Controlled heating of aluminum hydroxide results in a highly porous, granular form of aluminum oxide, Al_2O_3. Activated alumina shows preferential adsorptive capacity for water (hence, a good drying agent) and when saturated can easily be regenerated by heat. Besides being used as a desiccant for gases and vapors, other applications are as a catalyst, catalyst carrier, water purification, pigments, and abrasives.

Activated carbon, activated charcoal. A form of carbon that has (a) a porous or honeycomb-like structure and therefore a large surface area and (b) high adsorbtivity (certain molecules stick to it). Used to strip out impurities or extract selected compounds.

Activator. (1) An organic substance that makes possible the cross-linking, especially in vulcanization of rubber; (2) a fatty acid that increases the effectiveness of acidic organic accelerators.

Acyclics. Compounds having the structure of straight chains, not cyclic compounds.

Acylation. Same as acetylation.

Addition. A polymerization reaction in which monomers combine with one another to form a polymer without the formation of small by-products such as water.

Adiabatic change. Describes a process in which no heat is allowed to leave or enter the system.

Adsorption. The adherence of atoms, ions, or molecules of a gas or liquid to a solid substance called an adsorbent. Activated charcoal adsorbs odors and other contaminants from a variety of gases.

Aerobic. Usually refers to chemical reactions involving microbes that require oxygen in their environment to survive.

Aerosols. A suspension of liquid or solid particles in a gas. Aerosols generally refer to a packaging technique for gaseous products in sealed, pressurized containers. Aerosols consist of product, propellant, and the package. The contained product may be in the form of solution, emulsion, or suspension and can be dispersed by merely opening a valve. Propellants are typically CO_2, N_2O, and N_2. Typical uses include hair sprays, shaving soaps, paints, insect sprays, and deodorants.

Agglomeration. The tendency of large molecules or colloidal particles to bunch together, allowing removal from the solution, especially in polymerization process solutions.

Alcoholysis. A reaction between an alcohol and another organic compound that forms a new substance plus water. Typical examples include the reaction of alcohol (R-OH) with itself or another alcohol (R'-OH) to give an ether (R-O-R') and water.

Aldol condensation. A misnomer from the contraction of Aldehyde and alcohol, but alcohol is a subsequent step. Aldol condensation involves the reaction of an aldehyde with itself or another aldehyde (dimerizing) in the presence of an alkaline catalyst. The resulting dimer is also an aldehyde. Commercial production of 2-ethylhexanol involves the aldol condensation, dimerization of butyraldehyde, followed by the hydrogenation to 2-EH.

Aldox process. A two-step process combining the Oxo process and the Aldol Process. For example, propylene is reacted with carbon monoxide and hydrogen to form butyraldehyde (Oxo process), which is then dimerized and hydrogenated to give 2-ethylhexyl alcohol.

Alicyclic. Closed ring structures that fall into one of three different subgroups: (1) saturated cycloparaffins—also called naphthenes—such as cyclohexane or cyclopentane, and (2) cycloolefins such as cyclo-pentadiene—but not to be confused with aromatic compounds with the benzene ring.

Aliphatic. Straight- or branch-chain organic molecules that have saturated bonds (paraffins), double bonds (olefins), or triple bonds (acetylenes).

Alkali. Substances that, in water solution, have a high pH (they are basic and give up an OH⁻ ion) and more or less irritate the skin. Examples: sodium hydroxide, sodium carbonate, and the corresponding potassium compounds.

Alkane. A general term for a saturated hydrocarbon, C_nH_{2n+2}; a paraffin.

Alkene. A general term for an unsaturated hydrocarbon, C_nH_{2n}; an olefin.

Alkoxides, metal. These are also referred to as metal alcoholates. In these products, the hydrogen atom of the alcohol signature, -OH, is replaced by a metal or nonmetal such as sodium, potassium, magnesium, aluminum, titanium, boron, or silicon. These products are manufactured by the reaction of the metal with the alcohol to give metal alkoxides, with hydrogen as a by-product. End uses include catalysts, drying agents, fireproofing, and use as a cross-linking agent in polymerization reactions.

Alkoxylation. A reaction of alkoxides with an alkyl halide to produce ethers. (ex: R-O-Na + R'Cl = R-O-R' + NaCl).

Alkyds. The name comes from Alky(l) and (aci)d and refers to the thermoset resin coming from the reaction of a polybasic acid or anhydride such as phthalic acid or maleic anhydride and a polyhydric alcohol such as ethylene glycol.

Alkyl. A paraffinic hydrocarbon group derived by dropping a hydrogen atom from an alkane, C_nH_{2n+1}. For example, the methyl alkyl is CH_3-; ethyl is C_2H_5-. Groups such as this are often represented in the literature by the letter *R*-.

Alkylation. In petrochemicals, any reaction involving the thermal or catalytic addition of an olefin to a branch-chain hydrocarbon or aromatic hydrocarbon. The most notable example in petrochemicals is the addition of ethylene or propylene to benzene to produce ethylbenzene or isopropyl benzene (cumene). Other examples include the production of detergent alkylates.

Alkyne. An unsaturated hydrocarbon having one triple bond, C_nH_{2n-2}.

Allyl compounds. Highly reactive organic compounds containing the characteristic linkage $CH_2=CH-CH_2$-. Two common allyl compounds are allyl chloride and allyl alcohol. Both are colorless, pungent, mobile liquids that are used extensively as chemical intermediates in the manufacture of glycerol and epoxy resins. Allyl esters such as diallyl phthalate, diallyl iso-phthalate, and triallyl cyanurate are important monomers used to produce thermosetting molding resins and plastics with excellent mechanical properties, and solvent and heat resistance.

Alpha olefins. Straight-chain olefins with the double bond in the alpha position, i.e., in the first carbon-carbon bond of the chain. Alpha olefins start with the butenes and go up to the C_{30}'s and higher.

Alumina. Aluminum oxide (Al_2O_3). A white powder available in several commercial grades—technical, high purity, and CP. Used as a catalyst and catalyst support.

Amidation. Reactions leading to the formation of amides $R-\overset{\overset{\displaystyle O}{\displaystyle \|}}{C}-NH_2$. Examples include the dehydration (water removal) of the ammonium salts of carboxylic acids; reaction of ammonia with acyl chlorides, acid anhydrides, or esters.

Amides. Amides are produced by reaction of a carboxylic acid (fatty acid) or ester with anhydrous ammonia or amine (R'-NH_2.) Organic amides are characterized by the signature group—$CONH_2$.

Amination. A continuous process in which aliphatic and aromatic amines are produced by (1) high pressure, catalytic hydrogenation of nitro compounds (-NO_2 or nitriles (-CN)) and (2) action of ammonia on a chloro- or hydroxy-compound.

Amines. Aliphatic amines make up a class of organic compounds derived from ammonia (NH_3) where one, two, or three hydrogen atoms are replaced by alkyl groups. Amines are widely used as chemical intermediates and surfactants for fabric softeners, asphalt emulsifiers, petroleum additives, and ore-flotation agents.

Ammonolysis. Reactions involving ammonia. Ammonolysis of esters, acyl chlorides, and anhydrides give amides; aniline is produced by ammonolysis of chlorobenzene. The reaction is analogous to hydrolysis, with ammonia substituted for water.

Ammonoxidation. The contraction of two terms, ammonia and oxidation, ammonoxidation is a one-step process involving vapor phase, catalytic reaction of propylene with ammonia and the oxygen portion of air to give acrylonitrile. The ammonoxidation reaction is carried out at about 800°F and 30 psi.

Amorphous. Strictly speaking, characterized by noncrystalline structure, i.e., having no lattice structure that most solids have. Liquids are amorphous. Glass is a solid but is amorphous because it is considered a high viscosity liquid.

Amphoteric. Compounds having the capability of acting as an acid or a base. Amino acids are amphoteric—their molecules contain both an acid group (-$COOH$) and a basic group (-NH_2).

Anaerobic. A type of chemical reaction that occurs in the absence of oxygen such as the fermentation of sugars by yeast or decomposition of sewage sludge by microorganisms.

Angstrom. A measure of length used to denote distances between atoms and dimensions of molecules. Approximately 10^{-8} centimeter. The abbreviation is Å.

Anhydride. A chemical compound derived from an acid by elimination of the atoms that make up water. Example: SO_3, sulfur trioxide, is the anhydride of sulfuric acid, H_2SO_4. Anhydrides are a class of organic compounds derived from the combination of two carboxylic acids (R-COOH) by elimination of a molecule of water. For example, acetic acid (CH_3COOH) minus water gives acetic anhydride.

Anhydrous. An inorganic compound that does not have any water adsorbed on its surface or combined as water of crystallization. Not to be confused with anhydride.

Aniline ($C_6H_5NH_2$). Aniline is the largest volume organic amine and is the base material for many dyes and drugs. It is a colorless, oily liquid with a boiling point of 184°F. Aniline is produced by catalytic reduction of nitrobenzene ($C_6H_5NO_2$) or by reaction of chlorobenzene (C_6H_5Cl) with ammonia. Aniline is used as a chemical intermediate (organic synthesis), antioxidant, and rubber accelerator.

Anionic. Characterized by ions with a negative charge.

Antifreeze. Any compound or mixture that lowers the freeing point of water can be referred to as antifreeze. The preponderant, commercial antifreeze is ethylene glycol diluted with water. Other organic compounds used occasionally are methanol, ethanol, and propylene glycol. The primary application is to protect automotive cooling systems from freezing.

Antioxidant. Substances that retard or inhibit autoxidation at moderate temperatures and pressures. Commonly known, commercial antioxidants are aromatic amines, alkylated phenols, cresols, and hydroquinones.

Antioxidants are used as additives in plastics, rubber, gasoline, lubricating oils, and food products. *See* autoxidation.

Aprotic solvents. Type of solvent that does not accept or donate protons, such as benzene or dimethylformamide.

Aromatic. A broad class of unsaturated, cyclic, organic compounds characterized by the presence of at least one benzene ring.

Aryl. A group or radical derived from an aromatic hydrocarbon by the removal of one hydrogen atom from the ring structure.

Asymmetry. In the most common case, an organic compound having a carbon atom attached to four different atoms or groups is an asymmetrical atom.

Atactic. A characteristic of the spatial configuration of atoms or groups in a polymer chain. Atactic indicates a random distribution of those atoms or groups, i.e., no symmetry to the spatial configuration. This characteristic is important, for example, in determining the properties of polypropylene.

Atom. The hundred or so elements that make up matter are composed of particles called atoms. The atom is unique for each element and has consistent properties, including weight, number of neutrons, electrons, and protons. The atom is the smallest particle of an element that can exist alone or combine with similar or dissimilar atoms to form compounds (molecules) having unique characteristics.

Atomic number. The number of protons in the nucleus of an atom. The hundred or so known elements are usually arranged in the order of increasing atomic numbers for categorization purposes, with hydrogen (atomic number of one) the lightest, and uranium (atomic number of 92) one of the heaviest.

Atomic weight. The relative weights of atoms. The total mass of an atom is the sum of its number of protons, electrons, and neutrons. Hydrogen has an atomic weight of 2, carbon 12, oxygen 16, and sulfur 32.

Autoclave. A vessel used for heating, cooking, or sterilizing by exposure to superheated steam under pressure.

Autoxidation. Self-catalyzed oxidation in the presence of air. Autoxidation can be initiated by heat, light, or a catalyst. The commercial production of phenol and acetone from cumene is autoxidation. Other examples include the degradation of polymers exposed to sunlight for long periods of time; gum formation in lubricating oils and gasoline; and the spoilage of fats.

Avogadro's number. Amadeo Avogadro discovered one of the fundamental constants of chemistry, that equal volumes of gases at the same pressure and temperature contain the same number of molecules, regardless of the composition of the gas. The number of molecules in a volume of 22.4 liters of any gaseous element or compound is equal to Avogadro's number, 6×10^{23}. Avogadro's discovery allowed all sorts of determinations of molecular weights and relationships.

Azeotrope. A liquid mixture of two or more substances that behaves like a single substance, but most notably, the boiling point of the mixture is higher or lower than either of the substances. The vapor from boiling has the same composition as the azeotrope mixture.

B

Bactericide. Any substance that kills bacteria.

Bicyclic. An organic compound having two connecting ring structures that may or may not be the same type of ring. Naphthalene is bicyclic.

Bleaching agent. A compound used to whiten, lighten, or decolor a substrate by chemical reaction such as oxidation or reduction. Chlorine-containing bleaches include chlorine itself, sodium hypochlorite, sodium chlorites, and peroxides (hydrogen peroxide is the most common.) The reducing bleaching agents include sulfur-containing compounds such as sulfur dioxide, sulfurous acid, sodium bisulfite, sodium sulfite, and sodium hydrosulfite.

Boiling point. The temperature of a liquid at which its vapor pressure is equal to atmospheric pressure, allowing the liquid to easily vaporize with the addition of any heat.

Buffers. A substance in a solution that stabilizes the hydrogen ion concentration by neutralizing any added acid or base, such as a solution containing both a weak acid and its conjugate weak base, as with $CH_3COOH + CH_3COONa$.

C

Calendering. Passing a material (plastic, cloth) between rollers to make a thinner material or to give it a smooth finish.

Carbon black. Finely divided particles of carbon in powder form used in the manufacture of tires and other rubber products, plastics, inks and many other applications. Made by the incomplete combustion of oil or gas, usually the latter.

Carbonyl. The group $-C=O$, characteristically found in aldehydes, ketones, and organic acids. Also the general name given to compounds containing the CO group such as nickel carbonyl, $Ni(CO)_4$.

Carboxyl. The group -COOH, characteristic of the organic acids, such as acetic acid, CH_3COOH or adipic acid, $(CH_2)_4 (COOH)_2$.

Carboxylic acids. The organic acids which contain the group -COOH.

Catalysis. The phenomenon whereby a very small part of a substance can greatly accelerate the reaction of other substances while remaining unconsumed. Most industrial catalysis is performed by finely divided transition metals or their oxides.

Catalyst carrier. A neutral substance like alumina to which a catalyst can be bonded, mechanically or chemically, to permit proper exposure of the catalyst surface to the reactants.

Catalyst poisoning. The more or less permanent deactivation of a catalyst by chemical reaction with a contaminant. Sulfur will poison platinum catalysts; vanadium will poison zeolyte catalysts.

Catalyst promoter. A compound capable of either (1) activating catalysts at reaction temperatures; (2) reactivating spent catalysts; or (3) reacting with catalyst poisons.

Catalyst. Solids, liquids, or gases, which in small percentages accelerate or enable the reaction of two or more other chemicals without itself being consumed in the reaction. Sometimes the catalysts react momentarily with the other chemicals, creating conditions that permit the desired reaction to take place.

Cation. An ion having a positive charge, attracting negatives.

Celsius. Temperature scale in which $0°$ is associated with water freezing and $100°$ with water boiling at atmospheric conditions. (Originally the inventor, Andres Celsius designated the scale in the reverse direction but the Royal Society in London, in its wisdom, reversed it after he died.) *See* Fahrenheit for the rest of the story.

Chlorination. (1) Passing a liquid or gaseous stream of chlorine through or into a system. Used extensively in bleaching paper and pulp, treating swimming pools and municipal waste water, and processing some organic chemicals and metals; (2) the addition of chlorine to another molecule using molecular chlorine, Cl_2, or hydrochloric acid, HCl, or hydrochloric acid and oxygen using a catalyst, in which case it is called oxychlorination.

Colloid. Particles or bunches of particles that won't dissolve—they remain in suspension. They can be solid, gas, or liquid, and the medium in which they reside can be one of those forms.

Combustion. An exothermic chemical reaction involving oxidation of an organic compound and results in the creation of H_2O and CO_2. The heat results from rupture of the chemical bonds. In the case of organic materials such as wood, the original energy was stored by photosynthesis.

Condensation. (1) A chemical process in which two or more molecules combine and at the same time a separate small molecule such as water gets created. (2) The change of state from vapor to liquid, such as steam to water.

Condenser. A heat exchanger used to cool vapors to cause them to condense to liquid.

Coolant. Any liquid or gas capable of absorbing heat and transferring it elsewhere. Water makes an effective and cheap coolant.

Coordination compound. In polymerizations, a metal complex compound whose central atom provides a base for the monomer and the growing polymer to attach to as ligands.

Copolymerization. Polymerization of two or more dissimilar monomers such as the creation of SBR, rubber from styrene and butadiene.

Corrosion. The electromechanical degradation of metals or alloys due to the reaction with their environment. Acids or bases can accelerate corrosion. The resulting product is usually the corresponding metallic oxide, as in rust.

Critical temperature. The temperature above which a gas cannot be liquefied no matter the amount of pressure put on it.

Crystal. Despite the sloppy use in modern vernacular, crystal is the normal form of the solid state of any matter. Crystals have repeated characteristic shapes unique to a substance—some being more exotic and functional than others.

Curing. In the manufacture of thermosets, the interval necessary for extensive cross-linking (creation of bonds) to take place, usually by adding heat and/or chemicals.

Curing agent. A chemical added to a prepolymer that provides a site where cross-linking can take place later, transforming the prepolymer to a thermoset. For example, diethylene tetramine are used as curing agents for epoxy glue.

Cyanide. The -CN group, found in the deadly potassium cyanide or sodium cyanide crystals or vapors.

D

Dealkylation. The removal of an alkyl group (a straight chain) from a molecule. Sometimes the alkyl group is replaced by hydrogen, in which case the process is called hydrodealkylation.

Decarbonylation. The removal of the carbonyl group, -C=O, from a molecule.

Decarboxylation. The removal of the carboxyl group, -COOH, from a molecule.

Dehydration. The removal of all (stand-alone) water from a material by one of the following mechanical processes—heating, adsorption, absorption, centrifugal force, condensation, or flash drying.

Dehydrochlorination. The removal of chlorine and hydrogen from a molecule to form a chloride and water. Used, for example, as a step in a propylene-to-propylene oxide process.

Dehydrogenation. The removal of one or more hydrogen atoms from a molecule by chemical means, as in the conversion of alcohols to aldehydes. For example, methanol (CH_3OH) can be oxidized to formaldehyde (HCHO) plus H_2.

Density. The mass per unit volume. In the scientific and engineering community, mass is expressed in grams per cubic centimeter (g/cc) for solids and liquids and usually in grams per liter for gases. Regular people use pounds per cubic foot or pounds per gallon. The density of water is arbitrarily set as 1.0 g/cc.

Desiccant. A substance with large surface area per pound that will adsorb or absorb water vapor from the air. Popular desiccants are activated alumina, calcium chloride, silica gel, and zinc chloride.

Desorption. The process of removing the adsorbee from the absorber, the solid on which it is adsorbed.

Detergent. A substance that reduces the surface tension of water. It works at oil/water interfaces and has an emulsifying action, enabling it to remove dirt. It performs the same duties as soaps but is made from linear alkyl sulfonates, not the fatty acids that soaps are.

Detergent alcohol. The C_{12} to C_{18} range alcohols used to make detergents.

Dew point. The temperature at which air is saturated with water vapor; any temperature decline will cause the water vapor to condense into droplets or fog. More generally, the temperature at which any gas is saturated with a condensable component.

Diatomaceous earth. Material used as a filter or cleaning-up agent. Consists of soft bulky material composed of the skeletons of small, prehistoric, algae-like, aquatic plants (diatoms).

Diluent. A material added to another to reduce the concentration of a component or change a physical characteristic such as viscosity, reactivity, color, etc.

Dimer. A molecule formed by the union of two identical molecules. Hexene (C_6H_{12}) is the dimer of propylene (C_3H_6).

Dispersant. An agent added to a suspension solution to promote separation of the very fine particles, allowing or improving more complete chemical reaction.

Disproportionation. A chemical reaction in which a single compound serves as both an oxidizing and reducing agent such as the dealkylation of toluene to give benzene (the more reduced product) and xylene (the more oxidized product).

Dry ice. Solid carbon dioxide (CO_2). Gaseous CO_2 solidifies at $-110°F$ $(-79°C)$ without going through a liquid state.

Drying agent. A solid material, such as alumina, with a large surface area per unit weight that readily adsorbs moisture, i.e., moisture will collect on its surface. Used in gas drying towers and other applications that require the removal of moisture for quality control.

Drying oils. Naturally occurring oils used as paint binders or vehicles, usually of vegetable origin such as liquid triglycerides (naturally occurring esters) with three molecules of long-chain fatty acids attached to each glycerin molecule. Examples include linseed, tung, safflower, soya, tall oil, and cottonseed. Often used in outside architectural coatings because of their durability.

Drying towers. Vessels containing drying agents through which gases are passed to remove moisture. Drying towers usually come in pairs, allowing for continuous operations as the desiccant in one tower is purged of moisture using heat and a backwash of dried product while drying takes place in the other.

E

Elastomers. Polymers characterized by their flexibility and stretch, in particular various forms of rubber.

Electrolysis. Decomposition of a compound that is ionizable using electric current so that the positive ions go to the negative anode, and negative ions go to the positive cathode.

Electrolytes. A substance that will easily disassociate into ions when put into solution (generally water) and will conduct electricity. Sulfuric acid and sodium chloride are favorite media for electrolysis.

Emulsion and emulsifiers. An emulsion is a mixture of two liquids that do not mix very well—they are immiscible—but are held in stable suspension by the addition of a substance called an emulsifier. Emulsifiers are composed of molecules that attract the two liquids, one each at either end of the molecule, reducing the surface tension between the two liquids.

Endothermic. The characteristic of a chemical reaction indicating it absorbs or uses heat.

Epoxidation. The reaction in which an olefin is converted to a cyclic three-membered ether, i.e., an epoxide.

Epoxide. An organic compound containing the epoxy group (an oxygen and two carbon atoms connected in a triangular shape.)

Epoxy. (1) The name of a group consisting of an oxygen and two carbon atoms joined in a triangular shape; (2) the polymer formed by the reaction of a compound containing the epoxy group, often epichlorohydrin, and a diphenol, often Bisphenol A, to create a hard thermoset resin used as a bonding agent, coating, potting material, and many other applications.

Equilibrium. Forward and reverse reactions occurring at the same rate, resulting in a concentration of reactants. $A + B \rightleftarrows C + D$. Ammonia synthesis is an equilibrium reaction ($N_2 + 3H_2 \rightleftarrows 2NH_3$).

Ester. An organic compound derived from an acid where the active hydrogen located in the carboxyl group (-COOH) is replaced by an organic radical; the usual reaction involves an acid and an alcohol to form an ester (RCOOR').

Esterification. The reaction of an organic acid (having the signature group -COOH) with an organic alcohol (having the signature group -OH) to form an ester with the elimination of water.

Etherification. A chemical reaction forming ethers, such as the reaction of sodium or potassium alcoholates (RONa) with an alkyl halide (R'Cl) giving the ether ROR' and NaCl.

Exothermic. The characteristic of a chemical reaction indicating it gives off heat.

Extenders. Materials used in chemical formulations to add bulk or volume and reduce cost. Extenders range from materials like sawdust, glass fibers, and oils used to add bulk to phenolic resins to water held in a matrix, extending polyester resins.

F

Fahrenheit. The temperature scale invented by the 17th century instrument maker Daniel Garbriel Fahrenheit that has 32.2° as water's freezing point and 212° as the boiling point. (Originally Fahrenheit set the scale, thinking that 96° was normal body temperature and 32°, water's freezing temperature. Subsequently he realized that by his scale 0°F meant nothing to anyone so he cobbled up a solution of ice, water, and ammonium chloride that was in equilibrium at that temperature. But he never found any connection to 100° and didn't comment on water's boiling point. Unfortunately, the scale remains broadly used in America, to the chagrin of the scientific community.)

Fatty acids. Acids derived originally from animal fats or vegetable fat or oil containing a chain of alkyl groups with 4–22 carbon atoms, usually in even numbers, and ending with the carboxyl group, -COOH. The generic formula is RCH_2COOH, where R is an alkyl chain. Fatty acids can be saturated or unsaturated (olefinic, having a double bond).

Fatty alcohols. Alcohols originally derived from the corresponding fatty acids. The generic formula is $R(CH_2)OH$ where R is an alkyl group. Higher-weight fatty alcohols are also produced synthetically by the Ziegler Process or Oxo process. Alcohols in the carbon range of C_6–C_{18} are used primarily as plasticizers and detergents. They are often referred to as plasticizer alcohols (C_6–C_{10} alcohols) or detergent alcohols (C_{12}–C_{18} alcohols). The natural route to these alcohols involves hydrogenation of fatty acid esters from coconut oil, palm kernel oil, and tallow oil. The Oxo process involves reaction of an alpha olefin with synthesis gas (CO and H_2) to make an aldehyde (RCHO), which is then hydrogenated to give an alcohol. This process is used primarily for C_8–C_{12} alcohols. The Ziegler Process starts with triethyl aluminum, $Al(C_2H_5)_3$, as a root for adding ethylene molecules in series. Operating conditions determine the carbon chain length, C_8–C_{18} or even higher. To get the alcohol, the trialkyl aluminum is oxidized with air and then hydrolyzed with water. The Ziegler Process typically produces a distribution of about 50% C_6–C_{10} alcohols and 50% C_{12}–C_{18} alcohols.

Fatty esters. Esters made from fatty acids (RCOOH) by reaction with a linear alcohol (R'OH), generally methyl alcohol, giving the fatty ester (RCOOR').

Filter cake. The accumulation of removed matter upstream of the cloth or mesh in a filter drum used to remove solids or sludge from a liquid.

Filtration. Separation of solid matter from a liquid or gas stream by passing the mixture through a porous material such as sand, paper, cloth, or a membrane.

Fine chemicals. Chemicals produced in relatively small quantities and in a relatively pure state such as pharmaceuticals, perfumes, photographic chemicals, and reagent chemicals.

Fischer-Tropsch process (or synthesis) . The manufacture of synthesis gas (carbon monoxide and hydrogen) by passing steam over hot coal and the subsequent production of organic compounds from the synthesis gas.

Flash point. The temperature at which enough vapor forms above a liquid (or solid) to form an ignitable mixture with the air near the surface. The lower the flash temperature, the more hazardous the substance.

Flocculation. The process by which suspended solid particles aggregate in such a way that they resemble floccules, tufts of wool, which could impair the flow or appearance of the containing materials.

Fluidization. A finely divided solid is caused to act like a fluid by suspending it in a moving gas or liquid. Catalysts are often fluidized during reactions.

Foamed polymers. Thermosets and thermoplastics formed into low density, cellular materials containing bubbles of gas. Rigid foams have their gas bubbles in closed cells, inhibiting flexibility; flexible foams have the bubbles in open cells, permitting the gas to escape as the foam is flexed.

Free radicals. Free radicals occur when an atom group splits from a molecule. The molecular fragments then have one or more unpaired electrons. Consequently, radicals are usually short-lived, since they actively seek a satisfying mate. (Not to be confused with the term *group*, which is the *attached* radical). *See* radical.

G

Gas oil. A petroleum distillate with a boiling temperature range anywhere between about 450–800°F and is sometimes an olefins plant feedstock.

Gas-liquid chromatography. A process for determining the components of a gaseous stream. The gas is passed through a series of fine mesh screens. The finest are coated with a liquid. The setup causes the components to move through at different rates. At the tail end, they are detected, one after another, by thermal conductivity changes, density differences, or ionization detectors.

Gel. A jelly-like substance formed as solutes (materials in solution) from submicroscopic crystalline particle groups that capture much or all of the solvent in their interstices or structure. Usually, the solutes are a small proportion, maybe 1–2% of the gel. Gels can be used for filters and clarifying agents, textile and paper adhesives, polymerization precursors, and of course, as a component in *gelati*, the Italian ice cream.

Gum, arabic. This thickening agent used for stabilizing emulsions is made from the dried exudate of the stems of the *Acacia Senegal.*

Gypsum. Hydrated calcium sulfate ($CaSO_4 \cdot 2H_2O$). It occurs naturally and is used as a source of sulfur, as a drying agent, an organic liquid precursor, and in polishes.

H

Halogenation. Addition of one of the halogens to a chemical compound.

Halogens. One of the related chemicals, fluorine, chlorine, bromine, iodine, and astatine, listed in order of their decreasing chemical activity.

Hardening agents. Same as a curing agent.

Heat exchanger. A shell and tube type vessel allowing one hot liquid or vapor to transfer some of its heat to a cool liquid. E.g., a hot liquid runs through the tubes in a shell that has a cool liquid passing through it.

Heat of (combustion, crystallization, dilution, formation, fusion, hydration, reaction, solution, sublimation, transition, or vaporization). The heat required or absorbed or given up in each of these changes of state.

Heterocyclic. A closed-ring structure in which one or more of the atoms making up the ring (that usually has a five or six members) is something other than carbon such as nitrogen, oxygen, or sulfur.

Humectants. A substance such as glycerol with an affinity for water that is used to keep moisture content of products like baked goods and cigarettes in a narrow range.

Hydration. The reaction of water with another compound in which the H-OH bond is not split, as in gypsum ($CaSO_4 \cdot 2H_2O$). The result is called a hydrate and is written with a dot between the salt and the water.

Hydrazine. An extremely reactive compound, H_2NNH_2. Used in explosives and as a rocket propellant, precursors to polymer fibers, and as a polymerization catalyst.

Hydrodealkylation. The process of removing a methyl or larger alkyl group as in the conversion of toluene ($C_6H_5CH_3$) to benzene (C_6H_6). The alkyl group is replaced by a hydrogen atom.

Hydroforming. The use of hydrogen and pressure and a catalyst to convert an olefinic hydrocarbon to a branch-chain paraffin.

Hydroformylation (Oxo process). Production of aldehydes by catalytic reaction of an olefin with carbon monoxide and hydrogen. Cobalt catalysts are often used.

Hydrogenation. The chemical addition of hydrogen to another substance, usually an unsaturated organic compound, as opposed to hydrogenolysis.

Hydrogenolysis. The incidence of bond cleavage in an organic compound with the immediate addition of hydrogen to each dangling fragment.

Hydrolysis. A chemical reaction involving water such as a methyl ester reacting with water to form the corresponding acid plus alcohol or the conversion of natural fats to fatty acids and glycol with water.

Hydrophilic compounds. Materials that have a strong affinity for binding or absorbing water, sometimes forming gels.

Hydrophobic compounds. Materials that strongly repulse water such as of oils, carbon black, waxes, and many resins.

Hygroscopic. Substances capable of adsorbing moisture from the air, such as silica gel and calcium carbide.

I

Inhibitor. A compound that retards or stops undesired chemical reactions such as corrosion, oxidation, polymerization, or gum forming. Inhibitors are, more or less, negative catalysts.

Initiator. A compound used to start the polymerization of a monomer. Initiators are similar to catalysts, but they are consumed in the polymerization or reaction.

Ion. An atom or radical group that has lost (cation) or gained (anion) one or more electrons, thereby acquiring an electrical charge. Cations are positively charged ions and anions are negatively changed ions.

Isomer. Molecules having the same number and kind of atoms but differing in the spatial arrangement as in normal butane and isobutane.

Isomerization. A process used to convert straight-chain to branch-chain hydrocarbons as in a butane isomerization plant.

Isotactic polymer. A type of polymer structure in which groups of atoms that are not part of the polymer backbone are all in the same plane. The backbone groups are all in another plane.

Isotrope. Having the same atomic number (and position in the Periodic Table of Elements) but different masses. The difference is due to extra neutrons in the nucleus. For example hydrogen, one of three isotopes, has an atomic number of 1 and a mass of 1; the naturally occurring deuterium has a mass of 2 because it has an extra neutron in its nucleus; the artificially produced tritium has another neutron for a mass of three. All three have one proton and electron and, hence, an atomic number of 1.

K

Kelvin temperature scale. The scale used in theoretical physics and chemistry. Degrees Kelvin ($^{\circ}$K) is equal to degrees Centigrade ($^{\circ}$C) plus 273.

Kieselguhr. Same as diatomaceous earth.

L

Lachrymator. A substance that causes one's eyes to tear excessively.

Lacquer. A coating comprised of resins such as cellulose esters or ethers, shellac or gum, or alkyd resins plus a solvent such as ethyl alcohol that evaporates easily. The application process involves no chemical change, only the evaporation of the solvent leaving behind a hard, durable finish.

Latex. A milk-like fluid containing small particles of natural or synthetic rubber suspended in water. Synthetic latexes are made by carrying out a polymerization step in aqueous medium (an emulsion polymerization). Water-base paints are made from polyvinyl acetate latex created in this way.

Leaching. The process of extracting a soluble material by percolating water through it to dissolve the material. Later the water and dissolved material are separated by various means. Leaching is used to remove salts from ores.

Ligand. As it is used in polymerization, a ligand is a molecule or a group that attaches itself to a central atom of a catalyst in a complex and arcane way called a coordination compound. *See* Ziegler catalyst as an example.

M

Mass number. The number of neutrons and protons in the nucleus of an atom. The mass number of helium is 4, of carbon is 12, of oxygen is 16, and of sulfur is 32.

Merox. A proprietary catalytic process for extracting and converting mercaptans in naphtha and kerosene range hydrocarbons to reduce the sulfur content.

Metallocene. An organo-metallic coordination compound, or more specifically a cyclopentadienyl derivative of a transition metal or metal halide. Metallocenes are best known as catalysts for polymerizing ethylene and propylene.

Metathesis. A chemical reaction involving a double displacement. That is, groups from two different compounds play musical chairs and form two new compounds when the music stops.

Methanation. The catalytic reaction of CO_2 and/or CO with H_2 at 900–1000°F to form CH_4 and H_2O, as in a steam methane reformer.

Micelle. An electrically charged aggregation of large organic molecules in suspension (a colloid), typically in water if it's an emulsion polymerization process. The colloid is electrically charged and forms the site where polymerization takes place, even as the micelle stays in suspension.

Mineral oil . (1) Any liquid petroleum product; (2) a colorless, tasteless oil used as a laxative.

Mineral spirits. A refined heavy naphtha in the 310–380°F boiling range used as a solvent.

Miscibility. Similar to solubility. The ability of a liquid to mix completely with another. Bourbon is completely miscible with water.

Molality. A measure of concentration in a solution. A unit of molality is moles of solute per kilogram of solvent. The abbreviation is *m*.

Molarity. A measure of concentration, a unit of molarity, *M*, is equal to the molecular weight in grams of a substance dissolved in enough solvent to make one liter of solution. A 1-M solution of sodium hydroxide (NaOH) equals 40 grams of NaOH dissolved in 1 liter of water.

Mole. The amount of a pure substance equal in grams to its molecular weight.

Mole fraction. The ratio of the number of moles of one constituent of a mixture or solution to the number of moles of all the constituents.

Mole sieve. A substance with microscopic pores made of zeolytes or other compounds. The pores or channels in the substance are of atomic dimensions and will attract and allow the entrance of only certain size molecules. Mole sieves are therefore useful in the separation of smaller from larger size molecules of otherwise very similar characteristics such as boiling temperatures.

Molecular weight. The sum of the atomic weights of the atoms in a molecule. E.g., methane (CH_4) is 16.043, the sum of 12.011 plus 4 times 1.008.

Monomer. A molecule of relatively simple structure and low molecular weight that is capable of being polymerized with itself or other monomers into polymers, synthetic resins, or elastomers such as ethylene, propylene, styrene, butadiene, or vinyl chloride.

N

Naphthenes. Same as cycloparaffins.

Natta catalyst. A stereospecific catalyst made from metal alkyls and titanium chloride developed by the chemist Giulio Natta. *See* also Ziegler-Natta catalyst.

Neoprene. An elastomer with the repeating structure, $-C_4H_7Cl-$, that comes in the form of solid, latex, and flexible foam. Used as coatings, sealants, protective garments, adhesive tapes, etc.

Neutralization. A chemical reaction involving an acid and a base to give water and a neutral salt, as in $NaOH + HCl \rightleftarrows NaCl + H_2O$.

Neutron. Fundamental particle found in the nucleus of all elements except hydrogen. A neutron has a mass of 1.009 and no electrical charge.

Nitrile. An organic compound containing the -CN group. Examples are acrylonitrile (CH_2CHCN) and acetonitrile (CH_2CN), which have a triple bond between the carbon and nitrogen.

Noble metals. Not necessarily the metals with high price tags but rather the nine metals that have no or very little reactivity: gold, silver, platinum, palladium, iridium, rhenium, mercury, ruthenium, and osmium.

0

Oleophilic. Compounds that have a strong affinity for oils.

Oleophobic. Compounds that have a strong repulsion to oils.

Oligomer. A low molecular weight polymer consisting of a few repeating monomer units. Dimers, trimers, and tetramers are oligomers.

Oligomerization. The process of growing polymers of limited size—in the range of 2 to 20 or 30 or so repeating monomer units.

Organosol. A mixture of polymer and plasticizer used for molding. Organosol comes from the words organic and solvent. The polymer, which is the organic and is typically PVC, in fine particles is mixed as a colloid in plasticizer plus a little solvent. It all forms goo that can be put in a mold. With a little heat, the solvent evaporates, and the plasticizer and polymer-particle colloid form a gel, the final product.

Oxidants. The oxidizing agent. *See* oxidation.

Oxidation. While the term originally meant a reaction in which oxygen combined chemically with a compound, it eventually broadened to mean any reaction in which electrons are transferred. The substance that gains electrons is the oxidizing agent. Reductions always take place simultaneously, i.e., a reducing agent loses electrons. For example, if Fe, the iron metal, is oxidized by Cu^{++}, the cupric ion (it would be in a water solution), then the result is Fe^{++}, the ferrous ion plus the metal Cu, copper. Cu^{++} is the oxidizing agent and Fe the reducing agent.

Oxidation number. The number of electrons that must be added to an atom in its combined state to convert it to its elemental form. For example, in $CaCl_2$, the oxidation number of calcium is +2 and chlorine is -1.

Oxo process. *See* hydroformylation.

Ozone. An unstable, pale-blue gas, and a sibling of oxygen existing in the form of O_3. Ozone can be formed in the atmosphere by subtle electrical transfers, especially from pollutants or by shock from lightening. As an industrial gas, it is used as bleach and purification of water due to its reactivity.

P

Packed column. A distillation column filled with an inert material to enhance vapor/liquid contact. The packing can be beads, pellets, Raschig rings, metal chains, or specifically shaped devices such as saddles, helices, or rings.

Palm kernel oil. The oil from pressing palm kernels; contains triglycerides of stearic, myristic, oleic, palmitic, and lauric acids (the more common fatty acids) and is used in soap manufacture and as a dispersant and accelerator in polymerizations.

Palm oil. *See* palm kernel oil.

Peroxide. Compounds containing –O–O– linkage. Peroxides are extremely reactive and give up atomic oxygen readily. They are used industrially as oxidizing or bleaching agent.

pH. A scale indicating acidity or alkalinity, that is the extent to which a substance gives off H^+ ions or OH^- ions when placed in a specific concentration in water. Water, the neutral compound, is set by definition at a pH of 7. The scale on either side of 7 is exponential, so a pH of 1 is really acidic and a pH of 13 is really basic.

Phase separation. Conditions in which substances de-absorb themselves from their host, such as gases forming above liquids or solids forming in liquids.

Phenolics. A class of thermoset resins made by the condensation of phenol or phenol-containing compounds with aldehydes such as acetaldehyde or formaldehyde.

Plasticizers. Compounds added to high molecular weight polymers to give them flexibility, softness, and stretch. Plasticizers can be added mechanically at the compounding or shaping stage or chemically by copolymerization. For example, dioctyl phthalate is mechanically added to PVC; vinyl acetate is copolymerized with PVC. Plasticizer content can vary from 5–40%.

Polarization. The development of an electrical charge when in solution (with water.) Water, alcohol, and sulfuric acid are polar but most hydrocarbons are not. Some carboxyl groups and hydroxyl groups are polar. Polarization is important to the formation of emulsions and the actions of detergents.

Polyol. Alcohols having three or more -OH groups. Glycerols have three.

Precursor. Any compound that is amenable to direct conversion to another specific compound is a precursor.

Promoters. A material added in relatively small quantities that increases the effectiveness or activity of a catalyst.

Proton. A fundamental unit of matter having a positive charge and a mass number of 1. Protons and neutrons make up the nucleus of atoms.

Pyrolysis. Transformation of molecules by application of heat. Predominantly molecules are cleaved into smaller pieces, but cracking is often accompanied by isomerization of some and by reforming others into higher molecular weight molecules in the mad scramble that goes on during the short interval when cracking takes place.

R

Radical. The symbol $R-$ is used to represent an organic group such as $-CH_3$ and $-C_2H_5$ in a chemical formula. Free radicals are represented as $R\bullet$, as in $CH_3\bullet$ and $C_2H_5\bullet$. *See* free radical.

Raffinate. In extraction processes, the stream that has had the extracted material removed from it is called raffinate, in contrast to the other produced stream, the extract. Usually associated with aromatics extraction from naphtha streams.

Raney nickel. A form of nickel with a sponge-like structure.

Rare earth. One of a group of 15 chemically related elements: lanthanum, cerium, praseodymium, neodymium, promethium, samarium, europium, gadolinium, terbium, dysprosium, holmium, erbium, thulium, ytterbium, and lutetium.

Raschig rings (columns). Metal tubes a few inches long used as packing in distillation columns.

Rearrangement. A chemical reaction in which atoms of a compound recombine to form a new compound having the same molecular weight but different physical and chemical properties.

Rectification. The process of fractional distillation in which a portion of the condensed vapors are returned (reflux) to countercurrently contact rising vapors. Some of the vapors condense in the falling liquid, some of the liquid vaporizes in the rising vapor. Feed in a rectified column is usually introduced in the middle; the top half of the column where the reflux takes place is referred to as the rectifying section; the bottom half where reboil takes place is referred to as the stripping section.

Redox reaction. A reaction where both oxidation and reduction occur simultaneously. *See* oxidation.

Reducing agent. A reactant that loses electrons in a chemical reaction, as opposed to the oxidizing agent that gains electrons. *See* oxidation.

Reduction. *See* oxidation.

Reforming. In refining, a catalytic process in which naphtha molecules are cracked, rearranged, and/or recombined for the purpose of increasing the octane number of the naphtha. Reforming is also the process of converting hydrocarbons and steam to synthesis gas (carbon monoxide and hydrogen).

Release agents. *See* abherent.

Resin. Different meanings in different parts of the petrochemicals industry: (1) any thermoset polymer; (2) polymers of any kind, including plastics ready for fabrication; (3) plastics that have been molded or extruded; and the now obsolete (4) natural polymers such as shellac, rosin, frankincense, and myrrh.

S

Saponification. An aqueous solution of an alkali compound, like sodium or potassium hydroxide, reacts with an ester to form an alcohol and the salt of the acid corresponding to the ester. The process is typically carried out with the esters of natural fats, and then the salt formed is called a soap.

Scrubber. A packed column or fractionator used to separate components of a gas stream. The gas vapor passes up the column as a liquid descends. The liquid will selectively dissolve some of the components of the gas and then exit the bottom of the column for separation of the two. The vapors containing the components targeted to be removed exits the top of the column.

Sequestering agent. A substance that causes the formation of coordination complexes. *See* coordination complex.

Silica gel. Amorphous silica (silicon dioxide) used (1) as an absorbent, particularly for removing water from gases including refrigerants and from enclosed spaces such as in packaging; (2) as a catalyst carrier.

Sintering. The process of agglomerating metal powders to increase strength, conductivity, or density using heat and pressure but at temperatures below the melting point of the metal.

Sizing. Material used to increase or improve the stiffness, strength, smoothness, or weight of fibers, yarns, fabrics, paper, leather, and the like. Examples are starch, oils, gums, waxes, polymers, and silicones.

Soap. The reaction product of a fatty acid ester and a metal hydroxide, usually sodium hydroxide. Soap lowers the surface tension of water, permitting emulsification of soil-bearing fats if the soap is used for washing, of monomers in solution if the soap is used for emulsification in a polymerization process. *See* saponification.

Soda ash. Sodium bicarbonate that has been calcined (heated) to drive off any water.

Solute. A substance that has been completely dissolved in a solvent. The resulting mixture is a solution.

Solvent. A substance capable of dissolving another substance (solute). Solvents are either polar (permit transfer of ions) like water or nonpolar like most hydrocarbons.

Sorbent. A substance that absorbs, adsorbs, or otherwise entraps moisture, gas, or fluids.

Specific gravity. The density ratio of a substance to a reference substance. Water (1 g/cc) is the reference for solids and liquids and air (0.00129 g/cc at 32°F and 760 mm Hg) for gases.

Specific heat. The quantity of heat required to raise the temperature of a material by one degree per unit weight at standard conditions. Expressed as Btu/lb/°F or cal/g/°C.

Stabilizer. (1) A fractionating column used to remove light gases from a material that is otherwise liquid at ambient temperatures; (2) A compound capable of keeping another compound, mixture, or solution from changing its chemical nature; a stabilizer can slow down a chemical reaction, keep components in emulsion form, or keep particles in a colloidal suspension from precipitating.

Staple. A loosely defined unit of fiber, often in conjunction with its length, as in short staple vs. long staple. Staples of fiber are woven into threads and are contrasted with filament that is continuous, extruded lengths.

Stereochemistry. The field of organic chemistry devoted to three-dimensional spatial arrangements of molecules. Deals with stereoisomers, compounds having identical chemical formulas but different spatial arrangement of their atoms, such as geometric (cis/trans) isomers and optical (isotactic, atactic, and syndiotactic) isomers.

Stoichiometry. The measurement of reactants and products of a chemical reaction. Fundamentals rule that the combined weights of reactants will equal combined weights of products in reactions going to completion.

Stripping. Removal of the light gases from a liquid stream using a fractionator, evaporation, or passing steam or hot air through it.

Structural foams. Rigid foams made from thermosets. *See* polymer foams.

Sublimation. The passage of a substance from its solid state to vapor state without going through the intermediate liquid state. Carbon dioxide and camphor sublime.

Substitution. Replacement of one atom or group in a molecule with another. When chlorine substitution for a hydrogen atom in benzene takes place, the result is chlorobenzene.

Superheated steam. Steam heated above 212°F by subjecting water or steam to heat and pressure.

Supersaturation. A condition of having more solute dissolved in a solvent than normal conditions would allow, usually by an indirect route such as heating and dissolving, followed by cooling. Typically an unstable condition.

Surfactant. Any compound that reduces the surface tension (the electrical force that keeps them apart) between two liquids or between a liquid and a solid when it is dissolved in the liquid. Surfactants can be wetting agents, emulsifiers, or detergents.

Syndiotactic. The spatial configuration of a polymer in which the groups of atoms that are not part of the backbone of the polymer are arranged in a symmetrical pattern in a plane or planes other than the plane of the backbone.

Synthesis gas. A process feedstock consisting of carbon monoxide and hydrogen made by steam reforming or the partial oxidation of methane.

T

Terpolymers. A polymer made from three monomers such as ABS (acrylonitrile-benzene-styrene).

Thermoplastics. Polymers that can be resoftened by application of heat or pressure and can be dissolved in solvents.

Thermosets. Polymers that cannot be resoftened by application of heat or pressure because they have cross-linked, three-dimensional bonds. They are not soluble, and they decompose when enough heat is applied.

Thickening agent. A hydrophilic substance used to increase the viscosity of liquid mixtures and solutions and to aid in maintaining stability of their emulsifying properties.

Thinner. A solvent or diluent usually used to reduce viscosity.

Thixotropic. The ability of some colloidal gels to liquefy when subjected to shaking, vibration, or ultrasonic treatment and then to revert back to a gel or jelly-like form after time.

Transalkylation. A chemical reaction involving the movement of a group from one molecule to another molecule, dealkylating the first, alkylating the second. An example would be the reaction in which two molecules of toluene form benzene and xylene, which involves a quick two-step of a methyl group and a hydrogen between toluene molecules (also called disproportionation).

Tranesterification. The reaction between an ester and another compound to form another ester.

V

Valence. The number that represents how elements combine with each other to form molecules. The valence relates to the number of electrons floating in the outermost orbit of an atom and therefore the number of other atoms that it can share electrons with, which is to say, combine chemically with. Hydrogen has a valence of 1; oxygen has a valence of 2; so water has the formula H_2O.

Vinyl. The group $CH_2=CH-$.

Viscosity. A liquid's internal resistance to flow.

Vulcanization. The process of inducing cross-linking of polyisoprene (synthetic rubber) using sulfur and heat to increase its strength and elasticity.

W

Wetting agent. A material that enables water to better penetrate or cover the surface of another material by reducing the surface tension of the water, such as soap.

Z

Zeolyte. A widely used catalyst, originally from a naturally occurring hydrated silicate of aluminum and either sodium or calcium but mostly now a fabricated ion-exchange resin that can contain potassium and diverse groups of sulfonated organic compounds or resins, depending on the catalytic reaction desired.

Ziegler catalyst. First developed by Karl Ziegler in 1952, this class of catalyst (stereospecific) is widely used in polymerizations. A typical Ziegler catalyst is made of a metal base with alkyl groups attached. During a polymerization reaction, both the monomer and the active end of the growing chain will attach themselves (the term is *be complexed*) as ligands to the metal atom of the catalyst. This unique juxtaposition of monomer and growing chain results in a high degree of control over the shape of the resulting polymer. A typical Ziegler stereospecific catalyst is prepared by reacting titanium tetrachloride with an aluminum alkyl in a hydrocarbon solvent.

Ziegler-Natta catalyst. Giulio Natta developed a catalyst based on his work with Karl Ziegler for polymerizing vinyl monomers to give stereoregular, "tailored," three-dimensional chains. The catalyst is based on aluminum alkyls and $TiCl_4$ or other transition metal halides.

EXERCISE ANSWERS ⬡

"Putting on the spectacles of science,
in expectation of finding the answer to everything looked at,
signifies inner blindness."

The Voice of the Coyote
J. Frank Dobie, 1888–1964

Chapter 1

1. This is a mixed bag of correspondence:
   Paraffins have the formula C_nH_{2n+2}
   Olefins have the formula C_nH_{2n}.
   Aromatics, or at least benzene, have the formula C_nH_{2n-6}
   Saturates are paraffins
   Examples of unsaturates are the butylenes
   Examples of isomers are the three xylenes
   Examples of cyclics are benzene, xylene, and toluene.

2. There are 3 isomers of pentane:

   $CH_3-CH_2-CH_2-CH_2-CH_3$
   Normal pentane

   $CH_3-CH_2-CH-CH_3$
   |
   CH_3
   Isopentane

   CH_3
   |
   CH_3-C-CH_3
   |
   CH_3
   Neopentane

3. -C$_2$H$_5$. Only five hydrogens, not six. One hydrogen has to be dropped from the configuration to reflect the bond that connects the group to some other group.

4.

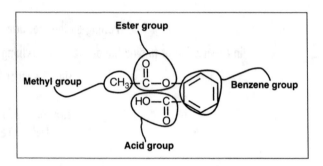

Chapter 2

1. naphtha - cat reformer - reformate
coal - destructive distillation - benzene
toluene - hydrodealkylation - benzene
gas oil - olefin plant - benzene
reformate - solvent extraction unit - benzene

2. Figure out the value of the pounds of toluene and hydrogen in, plus the operation cost, and compare it to the value of the benzene out:
   —500,000 times 7.21 times 0.20 = $721,000 for toluene.
   —from Table 2–2, you need 27 lbs. of hydrogen for every 1200 lbs of toluene, or 81,112 lbs.
   —81,112 times 0.40 is $32,445
   —total raw material cost is $753,445
   —benzene yield is 1000 lbs. for every 1200 lbs. of toluene, or 3,048,780 lbs.
   —3,048,780 times 0.24 = $731,703, which is less than the raw materials cost
   Answer is: sit tight and wait for the price of benzene to go up.

3. Platinum catalyst

4. Reformate, the product of reforming, has aromatics in it. Raffinate is the result of running reformate through an extraction unit, taking the aromatics out.

5. The feed is regular coffee, the solvent is CH_2Cl_2, the extract is caffeine, and the raffinate is decaf coffee. (Enjoy your next cup of raffinate.)

Chapter 3

1. Benzene, toluene, orth-xylene, meta-xylene, para-xylene, and ethylbenzene

2. toluene - hydrodealkylation - benzene
   mixed xylenes - adsorption - para-xylene
   toluene - disproportionation - mixed xylenes
   mixed xylenes - cryogenic distillation - para-xylene

3. Sure, because the freezing points of benzene, toluene, and meta-xylene are 167.2, 231.4, and 282.4°F, and far enough apart that distilling by freezing will work. The more usual method is fractional distillation, which is cheaper in both capital cost and operating cost.

4. e.

Chapter 4

1. When hydrogen atoms are added to the benzene molecule (hydrogenation), the chemical reaction gives off heat (exothermic).

2. a. six hydrogens
   b. catalysts
   c. nylon

3. You could pay almost 28.5 cents/lb:
   1000 @ 30 cents + 9 @ 0 cents = 944 @ X cents + 65 @ 40 cents
   + 1009 @ 0.5 cents
   X = 28.5 cents
   (the 1009 is the total feed, 944 + 65)

4. The difference between eggs and chickens and cyclohexane and benzene is that, "for sure," Nature made both cyclohexane and benzene at the same time.

Chapter 5

1. If you crack 1.25 billion pounds of propane or 2.778 billion pounds of gas oil, you'll get at least 500 million pounds of ethylene and more than the minimum amount of propylene:
   500 divided by 0.40 ethylene yield from propane
   = 1250 million pounds of propane feed
   200 divided by 0.18 propylene yield from propane
   = 1111 million pounds of propane feed
   500 divided by 0.18 ethylene yield from gas oil
   = 2777 million pounds of gas oil feed
   200 divided by 0.14 propylene yield from gas oil
   = 1428 million pounds of gas oil feed
   Cracking the propane gives 1.25 billion times .01 or 12.5 million pounds of butadiene; the gas oil gives 2.778 times 0.04, or 111.12 million pounds of butadiene.

2. Because it's inside a tube and does not come in physical contact with the fire in the furnace.

3. Hydrogen, methane, ethane, ethylene, acetylene, propane, propylene, butane, isobutane, butylenes (butene-1, butene-2, isobutylene), butadiene

4. For most of olefin plant history, there was plenty of propylene around, especially in refineries, so olefin plants didn't really have to be built to make propylene, only ethylene. But a petrochemical product that has

an annual U.S. demand approaching 20 billion pounds can hardly be considered second class.

5. Dehydrogenation of methanol, dehydrogenation of propane, metathesis of ethylene and butylene, and cat crackers. (Other crackers in refineries produce olefins too.)

Chapter 6

1. If they're in the chemical business and the price of butylenes is sufficiently lower than butadiene, which it often is, it makes sense to build a dehydrogenation plant to make butadiene from butylenes. But if they are a petroleum refiner and have no petrochemical business, then they want to get the butadiene out of the C_4 stream before it goes to the alkylation plant. Butadiene will cause the process to foam up, lowering conversion and making a general mess.

2. c. and d., depending on the price of butane and butylenes. In a. and b., butadiene is a by-product and the availability cannot be varied for all practical purposes. And Sal went into Chapter 11 last year.

3. Extractive distillation is used to remove butadiene from a C_4 stream; fractionation can be used to separate out butene-1; adsorption is also sometimes used to separate out butene-1; polymerization is sometimes used to pull out the isobutylene; dehydrogenation can be used to convert some of the butylenes and normal butane to butadiene; and alkylation is used to convert the butylenes to alkylate.

Chapter 7

1. Net cost of the cumene (benzene + propylene + operating cost):

681 times 1.60/gal. Divided by 7.19 lbs./gal	= 151.54
367 times 0.25 lb.	= 91.75
1048 times 0.20/lb.	= 209.60
	452.89

    452.89 divided by 1000 equals $0.453 per lb. of cumene.

    Net cost of the phenol (cumene + oxygen + plant operating cost less acetone less by-product values):

1000 times 0.453/lb.	= 453.0
300 times 0.500/lb.	= 150.0
725 times 0.25	= 181.25
-442 times 0.60	= -265.2
-99 times 0.10	= -9.9
	509.15

    Divide by 725 to get the breakeven cost of phenol. Add $0.02 to get the sales price of $0.722/lb.

2. Cutthroat forgot, once again, that when you produce all that by-product acetone (30 million lbs.) on the market, its market price is not likely to stay at $0.60/lb., and that changes the calculation of the breakeven cost of the phenol.

Chapter 8

1 ...cumene...styrene

2. Acetone could be considered as by- or coproduct of cumene manufacture. Propylene oxide is a by- or coproduct in one of the styrene processes.

3. | | |
    |---|---|
    | Benzene | 1000.0 lbs. |
    | to ethylbenzene | 1345.89 lbs. |
    | to styrene | 1188.95 lbs. |
    | using | 395.35 lbs. of ethylene |

Chapter 9

1. Quench is hot VC/HCL plus cool EDC. Oxychlorination is HCl, O_2, and C_2H_4 to EDC. Pyrolysis is EDC to VC and HCl. Chlorination is C_2H_4 and Cl_2 to EDC.

2. The addition of O_2 and Cl_2, as in the oxychlorination of EDC to styrene:
$$2CH_2=CH_2 + Cl_2 + \tfrac{1}{2}O_2 \rightarrow 2CH_2=CHCl + H_2O$$
or the oxychlorination of ethane to VC:
$$CH_3-CH_3 + Cl_2 + \tfrac{1}{2}O_2 \rightarrow CH_2=CHCl + HCL + H_2O$$

3. Chlorination is the addition of chlorine to a compound; oxidation is the addition of oxygen to a compound; oxychlorination is the introduction of them both, though the two chemicals may not end up in the same molecule at the end. Did you ever think you'd be comfortable with a word like oxychlorination?

Chapter 10

1.
epoxide	EO
silver bullet	silver oxide
epoxide ring	a molecule under stress
antifreeze ingredient	EG
EO by-product	CO_2
EG by-product	heavy glycols

2. The discovery and commercialization of the direct oxidation route using silver oxide as the catalyst.

3. From the material balance tables, 1.4 lbs. EO/lb. $C_2^=$ or 0.714 lbs. of $C_2^=$ required; also, 721 lbs. EO/1000 lbs. EG or 0.721 yield starting with 10 million gallons of EG:
   = 10 mmg EG
   = (10 x 9.3 lbs./gal.) mmlbs. of EG
   = (10 x 9.3 x .721) mmlbs. of EO
   = (10 x 9.3 x .721 x .714) mmlbs of ethylene required
   = 47.89 mmlbs. of ethylene required, theoretically

Chapter 11

1. epoxidation
   chlorohydrin...indirect oxidation...direct oxidation
   TBA...styrene...dipropylene glycol

2. Bad economics: chlorine disposal, energy intense

3. *See* Figure 8–8.

Chapter 12

1. The oxidation and the partial oxidation method, the CO_2 from an ammonia plant in a reaction with methane, and water all produce different ratios of CO and H_2. In addition, CO_2 can be removed by solvent extraction. So, the trick is to use two or three of these processes to get the $CO:H_2$ ratio to about 1:2.

2. Ammonia plants produce an otherwise worthless by-product, CO_2, which can be converted to CO and used as a co-feed to a methanol process.

3. Methanol is used primarily to make MTBE, acetic acid, and formaldehyde, and is used as a gasoline blending component or neat as an automotive fuel.

Chapter 13

1. Lot of arithmetic plus a little knowledge of chemistry. The reaction for MTBE calls for one mole of $iC_4^=$ and one mole of MeOH to make one mole of MTBE with a 98% yield. That calls for using the molecular weights, converting them to pounds and gallons, then applying the dollars. Using molecular weights to get the pounds:

$$56.1 \ iC_4^= \ + \ 32.04 \ MeOH \ \rightarrow \ (88.14)(0.98) \ MTBE$$

Converting to gallons and using dollars, where X = the breakeven value of the methanol:

($1.00/gal.) (56.1 lbs.) / (6.18 lbs./gal.)

+ ($X) (32.04 lbs.) / (6.59 lbs./gal.)

→ ($1.40) (86.39) / (6.18 lbs./gal.)

X = $ 2.02/gal.

Oh, and subtract the 10 cents operating cost.

2. Pros: aids the combustion of heavy ends in gasoline, it has a high octane number, and it has a low vapor pressure.

Cons: easily leaches out of leaked gasoline into groundwater or underground water, and some people complain the odor makes them nauseous.

Chapter 14

1. Because ethyl alcohol forms an azeotrope with water that is a constant boiling mixture, *i.e.* both the ethyl alcohol and the water, in a ratio of 95/5, boil together at a temperature different than either separately. Other examples in this chapter are the ternary azeotrope, ethyl alcohol - water - benzene; and DIPE - isopropyl alcohol - water. An azeotrope mentioned earlier is MEK - water - toluene raffinate used for toluene extraction.

2. Because "like dissolves like." The -OH group of IPA is a major part of the molecule, as it is in water, H-OH. That makes IPA "like" water. But in normal hexyl alcohol, $CH_3-CH_2-CH_2-CH_2-CH_2-CH_2-OH$, the -OH group is not a dominant part of the molecule, and "unlike" water.

3. Calculation as follows:

   Fermentation cost -

   $2.50/bushel divided by 2.6 gallons/bushel

   plus $0.50/bushel

   equals $1.46/gallon

   Petrochemical process -

   1 gallon of ethyl alcohol = 6.58 lbs.

   which requires X lbs. of ethylene to produce.

   1000/640 = 1.56 lbs. of ethyl alcohol per lb. of ethylene

   X = 6.58/1.56 = 4.22 lbs. ethylene needed to produce 1 gallon of
   ethyl alcohol

So, fermentation cost	$1.46/gal.
less plant operating cost	.30
gives	1.16
divided by 4.22 lbs. ethylene/gal.	
ethyl alcohol gives	$0.275/lb. ethylene

4.

	Direct hydration	Indirect hydration	Other process
Methanol	No	No	Yes
Ethanol	Yes	No	Yes
IPA	Yes	Yes	No
NBA	No	No	Yes
2-EH	No	No	Yes
SBA	Yes	Yes	-
TBA	Yes	Yes	-
BDO	No	No	Yes

Chapter 15

1. In the Ziegler process, long chains are grown on a root $(Al(CH_3)_3$, oxygen is added, and the whole thing is then clipped off to give the alcohol. In the Oxo process, long chain olefins are reacted with syngas

to make aldehydes that are then hydrogenated to the higher alcohols. (So that's 47 words. Who's counting?)

2. To get the long-chain olefins to feed the Oxo process, some producers use a Ziegler process to grow them on the $Al(C_2H_5)_3$ root. The difference then is only how to get from the $Al(C_2H_5)_3$–connected chain to the alcohol, via Oxo or via oxidation and hydrolysis.

Chapter 16

1. a. formalin
   b. formaldehyde
   c. formaldehyde...formaldehyde
   d. acetaldehyde or formaldehyde
   e. CO_2 or water
   f. aldehyde
   g. acetaldehyde

2. The dictionary (at least one late edition) says that an oenophile (*oeno-*, wine; *philos*, love) is a wine aficionado. And wine that has some aldehydes in it that never converted to ethyl alcohol is bad wine.

3. ethane → ethylene → ethyl alcohol
   ↓
   acetaldehyde
   ↓
   acetic acid
   ↓
   $CO_2 + H_2O$

Chapter 17

1. The only structural difference between the aldehydes and the ketones is in the position of the signature group, the double bonded oxygen. For ketones, it is always attached to a carbon in the middle of a hydrocarbon chain. For an aldehyde, it is always attached to a carbon at the end of the chain.

2. Using the material balances for an IPA plant, an acetone plane and an MIBK plant,
   1000 lbs. of MIBK needs 1160 lbs of acetone and 20 lbs. of hydrogen;
   1160 lbs. of acetone needs 1160 times 1.158=1343 lbs. of isopropyl alcohol
   1343 lbs. of isopropyl alcohol needs 1343 times 0.9=12090 lbs. of propylene
   So breakeven cost of the propylene is:

1000 times $0.45	450	
less hydrogen cost		
20 times $0.20	4	
less plant tolling costs		
1000 times $0.05	50	
1160 " "	50	
1343 " "	67	
equals	271	divide by 1209 lbs. of propylene
		breakeven cost of propylene=$0.224/lb.

3. a. formaldehyde   b. acetone
    c. formaldehyde   d. MEK
    e. nickel   f. zinc oxide
    g. isopropyl alcohol h. Wacker process
    i. methyl methacrylate j. butylene
    k. MEK   l. methyl isobutyl carbinol
    m. palladium chloride

Chapter 18

1.

   a. In an acid, the signature group has a hydroxyl group, -OH attached to the carboxyl carbon. In an ester, it is an -OR attached to the carbon where R can be an alkyl group like -CH$_3$, -C$_2$H$_5$, etc., but never a hydrogen atom.

b. Where the acid signature group has a hydroxyl attached to it, the aldehyde has a hydrogen attached instead.

c. An acid anhydride is two acid groups joined together by removing a water molecule, resulting in that wraparound signature group.

d. Oxidize an acid some more, and you end up with carbon dioxide (and water).

2. An acid has a double-bonded oxygen where an alcohol has two hydrogens.

3. Both of them can easily be converted to an anhydride.

4. Nylon 66...phenol...vinyl chloride...phthalic anhydride...styrene

5.

Starting Compound	Intermediate Compound	Final Compound
Ethyl alcohol	Acetaldehyde CH$_3$—CHO	Acetic anhydride
Ortho-xylene	Phthalic acid	Phthalic anhydride
Propylene	Isopropyl alcohol	Acetone
Cyclohexane	Cyclohexanone	Adipic acid

Chapter 19

1. Acrylonitrile and acrylic acid can both be made out of propylene. Acrylates and acrylic acid can both be made out of acrylonitrile.

2. In a fixed bed process, the feedstocks move over, around, or through the catalyst, which stays in one place all the time. In a fluidized bed process, the catalyst used behaves like a fluid as it moves through the reactor along with the feedstock. It is separated at the end of the reaction. Because the catalyst is so small and there is plenty of it, and because it gets surrounded by the feedstock, it effectively has a very large surface area per pond of catalyst. That facilitates the reaction, what with the feedstock coming into such intimate contact with all that catalyst. The major disadvantage of the fluidized bed reactor is the loss of the catalyst into the product stream during the catalyst recovery phase.

3. Vinyl chloride, vinyl alcohol, styrene, acrylic acids, acrylonitrile, the acrylates, propylene, acrolein.

4. Esterification starts with two compounds, such as an alcohol and an acid, and ends up with two compounds, an ester and water. The water is formed by each of the starting compounds giving up atoms to form the H_2O. Dehydration starts with one compound like alcohol and ends up with two, generally an olefin and a water. The water is made up of atoms given up by the original compound.

5. One of the olefins, propylene, is a commercial starting point for all those compounds. Commercial routes from ethylene, butylene, etc., have not been developed.

Chapter 20

1. Acetic anhydride (acetic acid), phthalic anhydride (ortho-xylene), and maleic anhydride (butane or benzene). Phthalic anhydride has a benzene ring as its unique feature, with its three double bonds; maleic anhydride has its single, but reactive, double bond. Acetic anhydride has only the

anhydride signature group. It has neither a closed-ring structure nor a carbon-carbon double bond.

The most reactive of the three is maleic because of the combination of the anhydride group and the highly reactive double bond.

2. Cobalt acetate is used to make acetic acid out of butane; palladium chloride/cupric chloride is used to make MEK out of butylene; and lo and behold, to make phthalic anhydride out of ortho-xylene or naphthalene, you use vanadium pentoxide.

3. The two most important factors are the cost of the feedstock and the yield of MA from each.

Chapter 21

1. In the higher alcohol process, the displacement is affected by oxidizing the trialkyl aluminum and then hydrolyzing, to form aluminum hydroxide and the linear alcohol. In the Ziegler process for alpha olefins, ethylene is used to displace the alpha olefin.

2. Because the addition is done by adding ethylene, the carbon count in that comes in two's.

3. Back in Chapter 3, Murphy's Law stated that anything that can happen will happen. The control of the rate of ethylene addition is not all that good. So, the rates vary from molecule to molecule and out comes a distribution of alpha olefins.

4. The Ziegler process produces the full range of alpha olefins, from C_4 to C_{20}^+. The Alpha Select process produces C_4 through C_{10}.

Chapter 22

1. Thermoplastics can be remolded several times by applying heat and/or pressure. Once thermosets are cured they cannot be changed in shape by heat or pressure.

2. Right, there's only three: initiation, propagation, and termination.

3. Bulk, suspension, solution, and emulsion polymerization.

4. Thermoplastics — because they need to be reoriented by pressure or temperature when they are drawn into fiber.

5. a. Homopolymer of formaldehyde (CH_2O)
   b. Copolymer of adipic acid and ethylene glycol
   c. Homopolymer of vinyl chloride

Chapter 23

1. When a polymerization uses two or more different types of monomers, it is called copolymerization. Examples are styrene acrylonitrile, LLDPE (ethylene and butene-1), and polyester (ethylene glycol and maleic anhydride).

2. The atactic polypropylene is soft and rubbery and doesn't have many useful characteristics.

3. Start with ethylene and chlorine and make ethylene dichloride; then crack it and make vinyl chloride; then polymerize it and make polyvinyl chloride.

Chapter 24

1. Typically, toluene diisocyanate and propylene glycol. Polyurethane can be either a thermoplastic or a thermoset depending on the monomer used. If propylene polyol is used in place of propylene glycol, there are more sites for cross-linking and a thermoset will result.

2. Rigid foams generally have closed cells; flexible foams have open cells so the air can escape during flexing.

3. Nylon isn't.

INDEX ⬯

"There is a time for many words,
and there is a time for sleep."

Homer, c. 700 b.c.

B

C

D

E

F

N

O

P

R

S

W

X

Y

Z